THE NATURE OF

ZAMBIA

A GUIDE TO CONSERVATION AND DEVELOPMENT ISSUES
No 2

The Nature of Zambia is the second volume in a series of information books about conservation and development issues around the world. Called The Nature of… series, it is not a profit-making venture, but forms part of the education and awareness work of IUCN's Conservation for Development Centre. This edition is based on the issues covered in Zambia's National Conservation Strategy (prepared by the Government of the Republic of Zambia, with assistance from the Conservation for Development Centre).

The Nature of Zambia was made possible by the generosity of Zambia Airways, Pamodzi Hotel, National Hotels Development Corporation, the World Wide Fund for Nature and the Royal Norwegian Ministry of Development Cooperation.

Conservation for Development Centre
IUCN
Avenue du Mont-Blanc
1196 Gland
Switzerland

ISBN 2-88032-403-3

Written and produced by Mark Carwardine

Design and illustrations © Christine Bass

Cover design by Roger Altass

Typeset by SP Typesetting, Birmingham and printed by Albany House Limited, Coleshill.

All the photographs were taken by David Reed/Impact (DR) with the following exceptions:

Steve Bass (SMJB): 44
Simon Bicknell (SB): 40, 47
Mark Carwardine (MC): 15, 17, 22, 41, 45, 47, 51, 55, 56, 64, 69
Chris Harvey (CH): 17, 18, 21, 22, 25, 32, 39, 56
Ian Murphy (IM): 10, 15, 17, 18, 19, 48, 50, 59, 60, 62

Further copies of titles in *The Nature of...* series are available from the Conservation for Development Centre at the above address.
Price: US$7.50/£5.00/Sfr15.00 (or equivalent) plus post and packing.
Copies are available in the countries of origin at cost price.

Other titles in this series:
The Nature of Pakistan

In preparation:
The Nature of Panama
The Nature of Saudi Arabia
The Nature of Oman
The Nature of Nepal
The Nature of Bangladesh
The Nature of Thailand

Founded in 1948, the International Union for Conservation of Nature and Natural Resources (IUCN) is the world's largest and most experienced alliance of active conservation authorities, agencies and interest groups. Its 500 members include States, government departments and most of the world's leading independent conservation organisations, in over 100 countries.

The Conservation for Development Centre (CDC) is the entrepreneurial arm of IUCN. An international, independent, non-profit centre, it specialises in the application of conservation principles to the process of economic development. Since its establishment, in 1981, CDC has become known worldwide for its work in ensuring that the exploitation of natural resources is done on a sustainable basis – to ultimately yield greater benefits for mankind – rather than for short-term gain. A small staff of under twenty, based in Gland, Switzerland, coordinates a worldwide network of consultants and experts for this work and is currently establishing CDC regional centres in a number of developing countries.

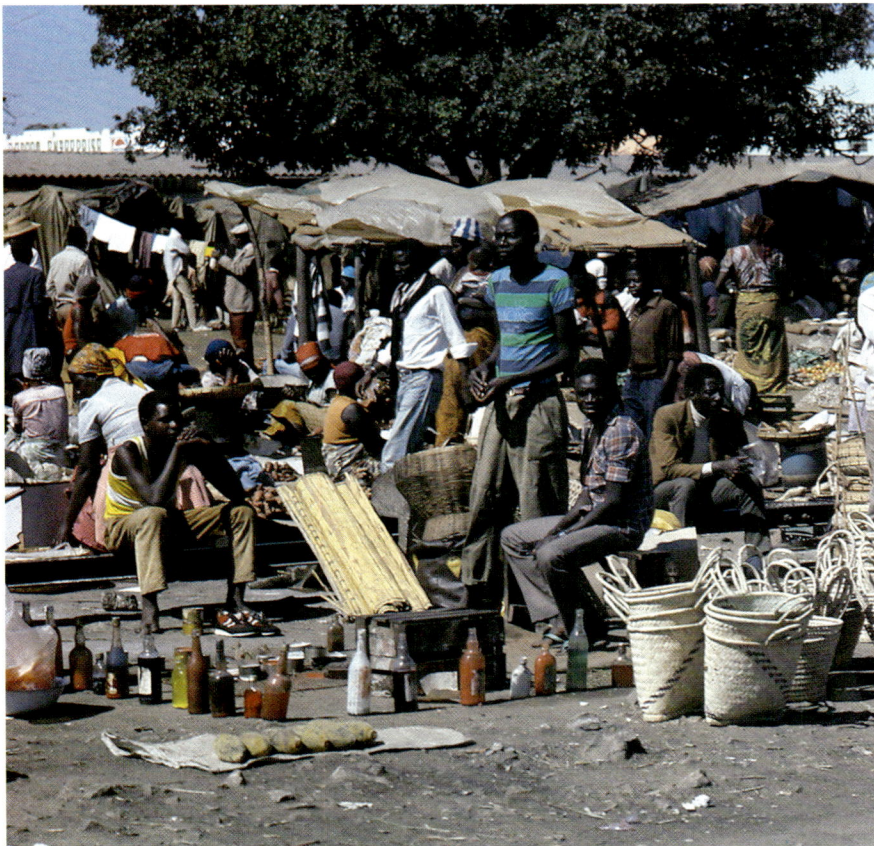

The World Wide Fund for Nature is an international conservation organisation with national offices on five continents. Since its establishment in 1961, WWF has set up and managed thousands of projects in over 135 countries and has brought its influence to bear on many critical conservation situations.

A Message

From The Hon. R.C. Kamanga, MCC

Since independence, the policy of the Party and its Government has emphasised the conservation, wise exploitation and development of Zambia's natural resources. The great thing about them is that – unlike our mineral resources – they are renewable over a relatively short period of time. Consequently, they are crucial to our future development.

Zambia recognises the need to diversify away from just mining and is therefore devoting considerable attention to developing the rich natural resources with which we are endowed. This means that conservation will be playing an increasingly important role in our country in the future.

In particular, we must step up our efforts to tackle the growing problems of deforestation, poaching, overfishing, soil erosion, destructive bush fires and pollution. These – and other – environmental problems are closely linked with Zambia's economic development and are a top priority of the Party and its Government.

They cannot, however, be dealt with in isolation. A more integrated and cross-sectoral approach to the conservation of natural resources is not just desirable but necessary if both conservation and development are to be sustainable. The acceptance by the Government of Zambia's National Conservation Strategy is testimony to the Party's commitment to this new approach to conservation.

A great deal of good conservation work has already been done in Zambia, by the Government and non-governmental organisations alike. But much more needs to be done in the future. With the help of the National Conservation Strategy – and despite the nation's economic problems – I am convinced that the future of conservation in Zambia is bright and that it will triumph over those that perpetuate the wanton destruction of our natural resources.

In his position as Chairman of the Rural Development Committee, The Hon. R.C. Kamanga, Member of the Central Committee, is responsible for the seven government departments which are concerned with the management and development of Zambia's natural resources.

Based on an interview with the author in 1986.

Contents

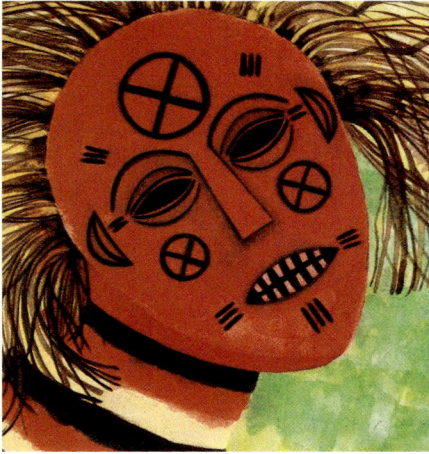

Chapter One

ZAMBIA: THE REAL AFRICA

*H*istory

Zambia's history can be traced back several thousand years, to the original small bushmen whose legacy of rock art is still being unveiled in many parts of the country. Invading Bantu tribes from the north forced the bushmen out and thus began a long, turbulent period in which Zambia was torn by tribal warfare and raids from surrounding areas. The turbulence continued well into the 19th century as Portuguese and Arab slave traders followed in large numbers.

Among the first Europeans to explore 'the golden heart of Africa' was the Scottish missionary, David Livingstone. His first expedition into the interior was in 1841, and he made many more over a period of nearly thirty years. It was in June 1852 that he embarked on the now-famous expedition during which he 'discovered' Victoria Falls. Livingstone spent many years in central and southern Africa, increasing the world's knowledge of the unknown continent and helping to put an end to slavery – which he referred to at the time as the 'great open sore of the world'. The last entry in his journal was written on 27 April 1873; four days later, on 1 May, he was found dead, kneeling by his bedside as if in prayer.

Less than twenty years after Livingstone's death, in 1889, Cecil Rhodes' British South Africa Company (BSAC) started to develop Zambia for mining purposes. At that time it was split into two regions, known respectively as North-Western Rhodesia and North-Eastern Rhodesia. Rhodes planned to extend British rule from the Cape to Cairo, but the Belgians in Belgian Congo and the Germans in Tanganyika foiled his master plan. The BSAC continued to administer both regions until they were amalgamated at the turn of the century, in 1911, to become the British protectorate of Northern Rhodesia. On 1st April 1924, power was assumed by the British Government through the Colonial Office.

The colonial regime continued until the creation of the Federation of Rhodesia and Nyasaland in 1953. Then the break-up of this Federation, only a decade later, paved the way for the establishment of Northern Rhodesia as an independent state.

Independence officially came on 24 October 1964, when the Union Jack was replaced by a green, orange, red and black flag, and a new name was taken from the Zambezi river, which flows across the south of the country.

Zambia has been ruled by President Dr Kenneth Kaunda, a strong supporter of Black nationalist movements in southern Africa, since independence. The President is leader of the ruling United National Independence Party (UNIP) – Zambia became a one-party nation on 13th December 1972 – and Commander-In-Chief of the Armed Forces. He appoints a Cabinet, led by the Prime Minister, to conduct the administration; the highest policy-making body, however, is UNIP's Central Committee, to which the Cabinet is subordinate.

Page from David Livingstone's sketchbook, showing remarkably accurate measurements of Victoria Falls; this was probably painted on his second and last visit there in September 1860.

But despite independence, in many ways Zambia remains a prisoner of its colonial past. In the 19th century, for example, its borders were marked on a map of Africa with no local consultation and a complete ignorance of relevant tribal factors. The King of Italy, acting as an adjudicator, actually drew a straight line for the western boundary with Angola. These artificial boundaries, of course, still persist.

Perhaps even more significant is the influence which comes from the discovery of huge copper deposits deep underground, when Zambia was destined to become a mining country. Everything else was largely forgotten or ignored and, to this day, few other nations in the world rely so heavily on a single commodity as Zambia does on copper. This reliance is now proving to be a serious hindrance to further development in the country.

Traditional dancing is still very much alive in Zambia.

Opposite, Top:
The Kuomboka Ceremony takes place in Western Province; it is an annual event celebrating the exodus of the Lozi chief and his people from flood-threatened land at the beginning of the wet season. (IM)

Bottom:
The motto 'One Zambia, One Nation' reflects the country's strong national identity and pride, irrespective of traditional ethnic groupings.

Culture and People

It is a tribute to the people and leadership of Zambia that their country remains a haven of peace in such a turbulent part of the African continent. Since independence, the previously rigid tribal barriers have gradually been broken down and President Kaunda's Humanism philosophy – which frowns upon the exploitation of man by man, supports human rights and condemns colonialism and racism – has helped to promote a gregarious and friendly atmosphere.

Today there is a motto – *One Zambia, One Nation* – which reflects the country's strong national identity and pride. There are still four major tribes and some 73 officially recognised ethnic groups but nowadays the people identify themselves first and foremost as Zambians, irrespective of other internal distinctions. Even marriages – which were once strictly within the tribe – commonly take place between people from different tribes.

However, the richness of tribal cultures remains distinct and traditional dancing is still very much alive in Zambia. A number of colourful cultural ceremonies also continue to be held, as they have since time immemorial. Perhaps the most spectacular of these is the Kuomboka, an annual event held during February or March when the Lozi people leave their homes for higher ground, as the rains swell the Zambezi; in ceremonial dress, and led by their chief in his striped royal barge, they move to their new homes (where they remain for the duration of the flood) then dance and sing the night away.

The total non-African population of Zambia – consisting mostly of Europeans and Asians – is about 60,000. The small communities of Europeans tend to work in mining and industry, and on large commercial farms; the Asians are mostly in commerce.

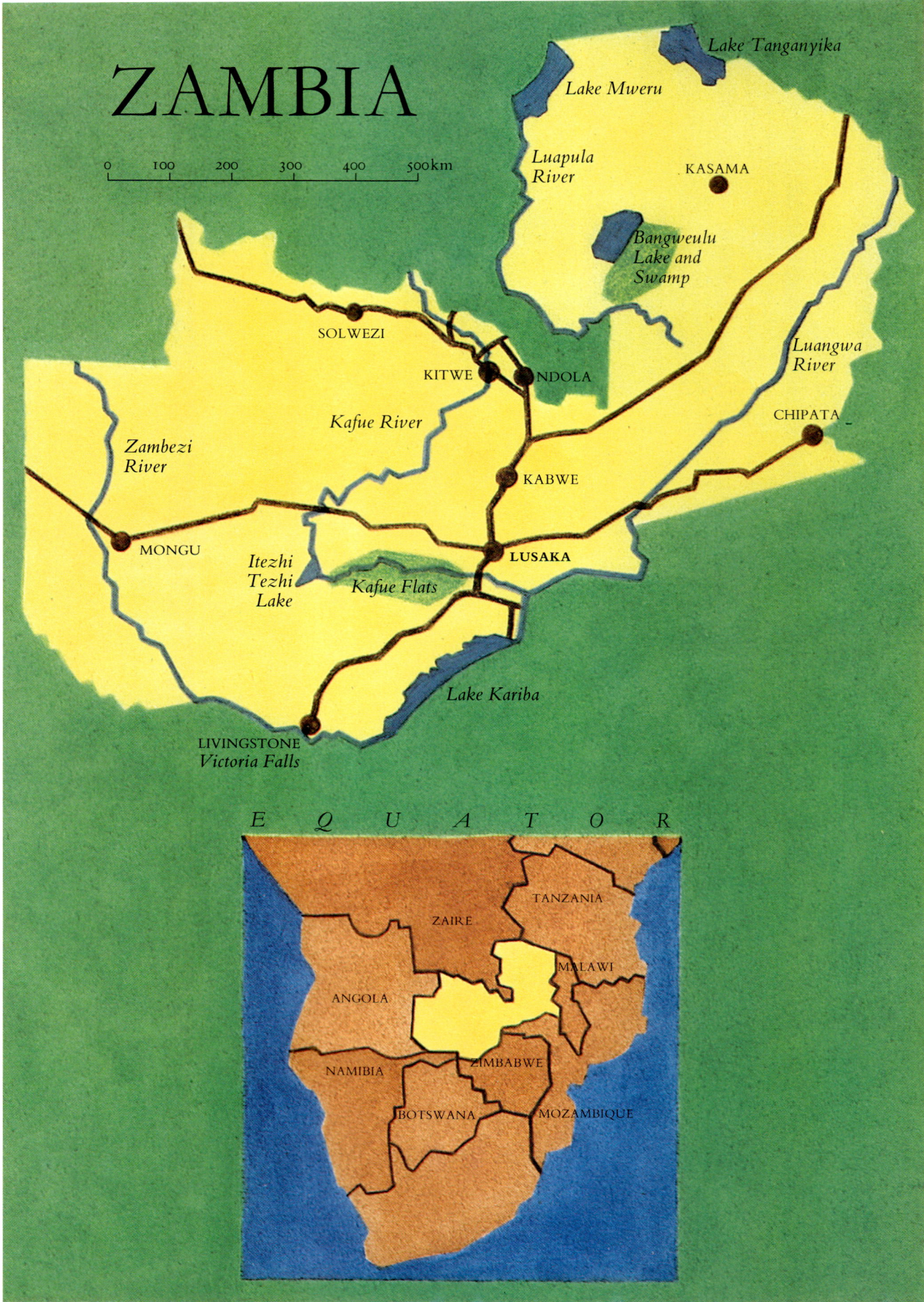

ZAMBIA

0   100   200   300   400   500km

Lake Tanganyika

Lake Mweru

Luapula
River

KASAMA

Bangweulu
Lake and
Swamp

SOLWEZI

Luangwa
River

KITWE   NDOLA

CHIPATA

Kafue River

Zambezi
River

KABWE

MONGU

Itezhi
Tezhi
Lake

Kafue Flats

LUSAKA

Lake Kariba

LIVINGSTONE
Victoria Falls

E Q U A T O R

ZAIRE

TANZANIA

ANGOLA

MALAWI

NAMIBIA

ZIMBABWE

BOTSWANA

MOZAMBIQUE

Zambia at a Glance

Total area:	290,586 sq mls (752,972 sq km);
Neighbouring countries:	Zaire, Tanzania, Malawi, Mozambique, Zimbabwe, Botswana, Namibia, Angola;
Population:	6,242,000 (mid-1982) with 3.1 per cent annual growth rate;
Urban population:	43 per cent (two-thirds in the Copperbelt) with 5.5 per cent annual growth rate;
Major towns:	Lusaka (capital), Livingstone (capital until 1935), Kitwe, Ndola, Chingola;
Languages:	English (official); over 80 other languages have been identified (notably Bemba, Nyanja and Tonga, together spoken by two-thirds of the population);
Religion:	Complete freedom of worship;
Government:	Zambia is a one-party participatory democracy, headed by the President, Dr. Kenneth D. Kaunda, who has been in office since independence on 24 October 1964;
GNP per capita:	US$500 (1979) with 0.8 per cent average annual growth rate (1960-79);
Labour force:	68 per cent in agriculture; 11 per cent in industry; 21 per cent in services (1979).

Geography

Zambia is a large, landlocked country with comparatively few people. The size of Denmark, France, Switzerland, Belgium and Austria combined, its population is less than 10 per cent of the total for all these European nations together.

Since nearly half all Zambians live in towns and cities – it is one of the most urbanised countries in Africa – there are vast areas left almost unpopulated. Even the national average is only about eight people per square kilometre. The population, however, is growing rapidly.

Transport routes and the location of minerals have greatly influenced Zambia's development. Almost all development since the mines' railway was built in the early 1900s – from the industrial towns of the north to the commercial farms of the south – has centred on this north-south line.

There is also a high density of much smaller settlements in the Mozambique and Malawi border areas, in the Bangweulu-Luapula complex of swamps, certain islands in the northern region and in the flats of the Zambezi river in Western Province. Populations are booming in several provincial towns, such as Mpika, due to the increased development of small-scale industries and aided by the Government's policy of decentralisation.

Most of Zambia lies on a high savanna-covered plateau, lying between 1,000 and 1,600 metres above sea level. The highest parts of the country are in the north-east, with the plateau gradually sloping to the south-west. This altitude has a moderating effect on what would otherwise be a harsh tropical climate, earning Zambia the nickname 'the air-conditioned state'. There are three main seasons: cool and dry (May-August), hot and dry (September-November) and warm and wet (December-April). October is the hottest month (when temperatures can reach nearly 40 degrees centigrade) and June and July the coldest; every few years somebody claims to see snow falling, but they are never quite believed.

About 70 per cent of the country is in the miombo woodland zone. Miombo woodland consists of an open mixture of shrubs, a variety of trees of moderate height and tall grasses. Several other varieties of woodland, forest and grassland also exist, their presence depending mostly on altitude and rainfall.

Zambia's main drainage systems are the Zambezi river (whose major tributaries are the Kafue and Luangwa rivers) and the Chambeshi watershed. There are also several large lakes – notably Mweru and Bangweulu – and part of Lake Tanganyika in the north.

Near Livingstone, the Zambezi river passes over the world-famous Victoria Falls. The explorer David Livingstone, who in 1855 was the first European to witness this awesome cataract, named the Falls in honour of his queen. He later described them as 'the most wonderful sight I had witnessed in Africa'. There are a few higher falls around the world, but none could equal Victoria in grandeur. At the height of the floods over five million litres of water plunges over the mile-long edge *every second* to the chasm 100 metres below.

Like generations of Africans before him – and countless tourists ever since – Livingstone first became aware of the Falls long before he was able to see them. The noise sounds like thunder and the clouds of spray, which billow hundreds of feet into the air, look like smoke rising from an enormous fire. The local name – Mosi-Oa-Tunya – actually means *the smoke that thunders*. The popularity of the Falls is such that the nearby small town of Livingstone, close to the border with Zimbabwe, has become the country's tourist capital.

Livingstone was once also the true capital of Zambia, but relinquished its position to Lusaka in 1935. Lusaka is a typical African city with broad, tree-lined streets, a bustling market and a mushrooming population. Most other major towns in the country are located in the mineral-producing region known as the Copperbelt, an area some 50 kilometres wide and 110 kilometres long near the border with Zaire.

Economics

The Zambian economy is currently in worse shape than it has ever been before. Since 1975 the country has been reduced to virtual bankruptcy – almost entirely because of its overwhelming dependence on copper. The land 'born with a copper spoon in its mouth' is no longer secure in the hands of the miners alone.

Between independence and the late 1970s, the Zambian economy was booming. This growth, however, reflected increased copper output and rising prices on the world market, rather than a general growth in all sectors of the economy.

Instead of building a secure and balanced economy, little thought had been given to anything but the immediate profitable export of copper, with the result that Zambia's economy was resilient only as long as the copper market held well.

But a series of events – totally outside Zambia's control – have had a disastrous effect on this market and, thereby, on the Zambian economy:
- copper production itself is suffering from a deterioration in the grade of ore, so the mines probably have only another 15-20 years of economic life;
- low rates of growth in the Western world, coupled with changes in electronic and communications technology have significantly reduced international demand for copper;
- consequently, Zambia's foreign exchange earnings have fallen sharply, forcing its external debt to some US$4.5 billion (1984);
- this means that spare parts and other equipment can be imported only in

Left:
Zambia is one of the most urbanised countries on the continent; nearly ten per cent of the population lives here in Lusaka. (MC)

Below:
In many ways, Zambia remains a prisoner of its colonial past. When huge copper deposits were discovered deep underground it was immediately destined to become a mining country and no effort was made to broaden its economic base. This miner is drilling holes for blasting, at the Mufulira mine. (IM)

Left:
Zambia's reliance on copper (for over 90 per cent of its foreign exchange earnings) constitutes one of the highest levels of dependence of any country on just one commodity. This has become a severe constraint to development. (IM)

The Government is diversifying Zambia's economic base away from mining; its current development aims are strongly biased towards agriculture.

small numbers, hence industry is operating well below capacity (30 per cent is average); efficient agriculture is also affected, with only 38 per cent of the country's tractors currently operational;

– the lack of industrial activity (which is heavily dependent on imports) has contributed to a reduction in government revenue, which itself has fallen by 30 per cent in the last ten years; consequently, expenditure on health is now 60 per cent of what it was ten years ago, while in education 20 per cent less in real terms is spent on each Zambian child.

To add insult to injury, cobalt – traditionally Zambia's second largest foreign exchange earner – has also been sliding in price. Again, the reason is lack of demand in the industrialised world.

There is clearly a desperate need to diversify Zambia's economic base away from mining. The Government's current development aims are strongly biased towards agriculture. But management of natural resources is generally poor, there is a serious shortage of agriculturalists and essential equipment is lacking. Consequently, large quantities of food still need to be imported every year – which is an unnecessary drain on foreign currency – when Zambia could be producing such basic supplies itself.

The Government has also embarked on an ambitious programme to utilise two other great assets – the prolific wildlife and Victoria Falls – which are expected to form the basis of a substantial tourist industry in the near future. The 1980s have witnessed growing numbers of visitors to Zambia – from all over the world – and they are becoming an important source of foreign exchange earnings.

Chapter Two

OFF THE BEATEN TRACK

A *Wealth of Wildlife*

Zambia's wildlife is among the richest in Africa. It includes several species which are found nowhere else in the world and enormous numbers of many commoner varieties.

There are more than 100 different mammals in the country, including distinctive 'big-game' species such as hippo, elephant, black rhino, lion and buffalo. Zambia is also the *only* place to see Thornicroft's giraffe, which is unique to the Luangwa Valley, and one of the best places in Africa for seeing leopard. It is also noteworthy for harbouring all three sub-species of lechwe: the black, which is confined to south-eastern Zaire and the Bangweulu swamps; the Kafue, confined to the central Kafue flats; and the red, which is still present in Botswana but more common in central and western Zambia.

Zambia's wildlife is among the richest in Africa and includes more than 100 species of mammals, 700 birds and 4,600 plants. The fish eagle is Zambia's national symbol. (Left to right: MC, DR, IM, CH, IM, CH).

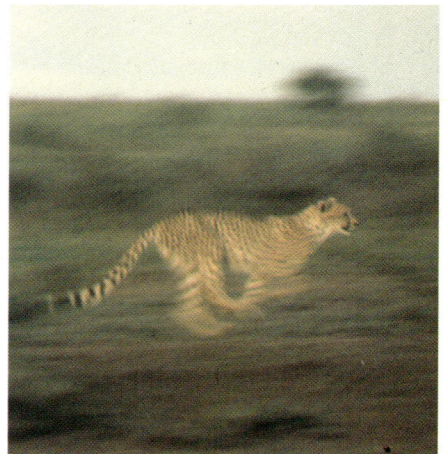

Above, left to right:
The serval is most active in the early morning or late afternoon, resting in the shade during the heat of midday. It is a fairly secretive animal and rarely seen. (CH)

The bat-eared fox spends hours on end listening for rustlings in the grass of termites and dung beetles, which are its favourite food. (IM)

The cheetah is the fastest of all land animals, able to reach about 100 km/h over short distances. (CH)

IUCN lists six species of threatened mammals as occurring in Zambia: wild dog, cheetah, black rhino, elephant, leopard and lechwe. There are also two threatened reptiles – the Nile and slender-snouted crocodiles – and several threatened birds, notably the shoebill, wattled crane and slaty egret. More than 700 different birds occur in the country altogether, including Zambia's national emblem, the fish eagle.

Some 4,600 different species of plants have so far been identified, more than 200 of which are found nowhere else in the world.

More than 700 different birds occur
in Zambia. Here, a malachite king-
fisher waits patiently for fish. (IM)

Opposite, Top:
Bird skins, collected from all over
Zambia, on display in a research
laboratory at Livingstone Museum.

Snake Man

Alick Chanda, snake expert and Curator of Munda Wanga Zoological and Botanical Gardens, has been bitten by snakes 89 times. Puff adders have given him most trouble, with no fewer than 23 bites in recent years; on one occasion, the bite was so bad that he was pronounced dead. But he rarely ends up in hospital after being bitten these days, preferring to treat himself at home.

Despite the hazards of handling snakes, Alick believes he could not live without them around him. He always likes to carry a house snake and a sand snake in his pockets wherever he goes; the sand snake once escaped on a flight to London, but these days he is apparently a little more careful.

Although Alick has been bitten by everything from gaboon vipers and boomslangs to black mambas and cobras, he hates to see anybody kill a snake. In all the times he has been bitten, with one exception, he maintains

that it has been his own fault. Most snakes will quickly get out of the way as soon as they feel the vibrations of anyone approaching. A quarter of Zambia's snakes are venomous – including some very common and deadly species such as gaboon vipers, forest and spitting cobras, boomslangs, night adders and black mambas – but very few cause much trouble in the wild. Puff adders are perhaps the only exception; they always stay put and are therefore easily trodden on. Since they are the commonest venomous snakes in Zambia, they are also responsible for most bites.

But even puff adders have an important role to play in Zambian ecology: as rat catchers. Zambia already loses enormous quantities of stored grain to rats but, without snakes, the loss would be far greater.

Snakes are also contributing to human welfare. Their venom is a crucial ingredient of many modern drugs, including some used in the treatment of cancer, arthritis, thrombosis and several other disorders. Who knows what other uses may become apparent in the years to come? Already, the venom of some species is extremely valuable and the most potent ones are highly sought after by drug manufacturers. Several African countries now earn a significant proportion of their foreign exchange earnings through snake venom sales – and the beauty is that the snakes do not have to be killed.

Zambia is very rich in snakes, with 27 species known to occur in the country altogether. Alick Chanda's attention has most recently turned to the Tanganyika water cobra. Lake Tanganyika is the only place in the world this species can be found. Although its venom is very potent – designed to kill fast-moving freshwater fish within seconds – it is a docile creature and causes no problem to the fishermen in the area. But in a forthcoming expedition Alick is likely to be handling many water cobras and, in his own words, who knows, the ninetieth bite may well be his last.

Far Left:
The forest cobra is one of 27 different snakes known to occur in Zambia. Although potentially one of the most dangerous, it will normally try to escape as soon as it feels the vibrations of someone approaching.

Left:
The boomslang's fangs normally remain folded back along the roof of its mouth. They drop down as the snake's jaws open in preparation for biting. Although one of the deadliest snakes in the world, the boomslang is a shy animal and rarely seen. (CH)

Opposite:
Alick Chanda has been bitten by snakes 89 times – once so badly that he was pronounced dead – but hates to see a snake being killed. He believes they have an important role to play as rat catchers and in medical research.

Zambia is the only place to see the Thornicroft's giraffe, which is unique to the Luangwa Valley. (MC)

Top:
Roughly the size of The Gambia, Kafue is the largest of Zambia's 19 national parks; together, they cover more than eight per cent of the country. (CH)

Protected Areas

Zambia contains some of the wildest and most beautiful wildlife sanctuaries in the whole of Africa. Several of its national parks – in particular, Kafue and South Luangwa – are known throughout the world.

Zambia's relatively small human population, which is to a large extent urban, has made an extensive system of protected areas possible. There are 19 national parks – compared with just one before independence – and a further 31 game management areas. In practice, control over these is minimal but, together, they put more than one-third of the country under some form of wildlife management.

Kafue National Park, to the west of Lusaka, is 22,500 square kilometres in area and one of the largest wildlife sanctuaries in Africa. Established in 1951, it consists of a vast and gently undulating plateau watered by two tributaries of the Kafue river, the Lafupa and the Lunga. These waterways are its lifeblood and flow directly into the Park from the north. The vegetation consists of open grassy plains, or 'dambos', with areas of miombo and mopane woodland and patches of Zambezi teak. Kafue is the best park in Zambia for cheetah and has abundant roan and sable. It is also a bird-watcher's paradise. Over 400 different species have been recorded around the Kafue Flats area alone, some of them occurring in quite large numbers.

Lying along the Luangwa river, a tributary of the Zambezi to the east of Lusaka, South Luangwa is a beautiful woodland Park with a wide selection of trees, including mopane, marula (whose fruit is supposed to make elephants slightly tipsy), winterthorn, sausage and baobab. It is best known for the abundance and variety of its large mammals. Hippo, buffalo,

ZAMBIA
WILDLIFE AREAS

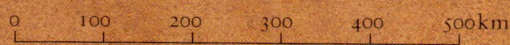

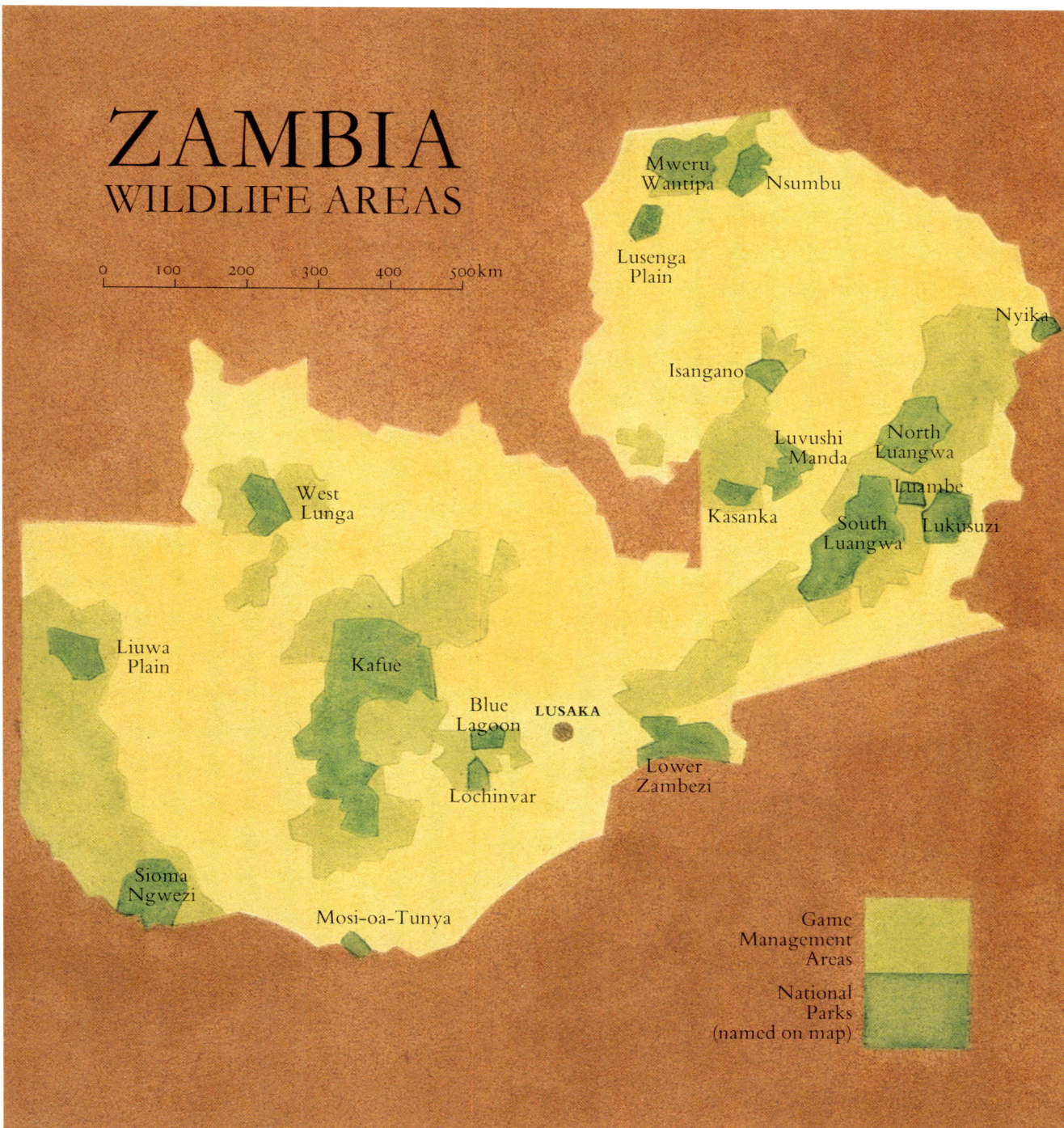

0   100   200   300   400   500km

Mweru Wantipa
Nsumbu
Lusenga Plain
Nyika
Isangano
Luvushi Manda
North Luangwa
Luambe
West Lunga
Kasanka
South Luangwa
Lukusuzi
Liuwa Plain
Kafue
Blue Lagoon
LUSAKA
Lochinvar
Lower Zambezi
Sioma Ngwezi
Mosi-oa-Tunya

Game Management Areas

National Parks (named on map)

elephant, impala, greater kudu, puku, sable and roan antelopes, wild dog, cheetah, leopard, lion, hyena, chacma baboon and many others are regularly seen in the Park. Local specialities include Thornicroft's giraffe and Cookson's wildebeest; black rhinos are also present, albeit surviving only in very small numbers. Less than half the size of Kafue – though still vast by any standard – South Luangwa is widely considered to be one of the finest wildlife sanctuaries anywhere on earth.

Other national parks of note include Lochinvar, devoted to the conservation of the red lechwe; Blue Lagoon, which is important for the Kafue lechwe and as a feeding ground for waterfowl; and Mosi-Oa-Tunya, occupying the northern bank of the Zambezi river and half the Victoria Falls.

M. N. Katanekwa

National Monuments

Waterfalls, gorges, extraordinary trees, old buildings, fossil forests, geological sites and many other natural and historical relics are all the responsibility of Zambia's National Monuments Commission. Created in 1947, the Commission is one of the oldest statutory bodies in the country.

The current Director is M.N. Katanekwa, an archaeologist by training and former Curator of Livingstone Museum. He sees the Commission's main functions as preserving, protecting and managing Zambia's natural and cultural heritage; presenting this heritage to the public, by providing facilities (such as the Field Museum at Victoria Falls) and through education and awareness programmes; and researching into Zambia's archaeological and palaeontological past.

There are currently 71 National Monuments around the country, some of them internationally famous. The first to be declared – soon after the Commission came into existence – was Victoria Falls, in 1948. The Falls are shared with Zimbabwe and there may soon be a joint effort – in Zambia, largely Katanekwa's responsibility – to make it a World Heritage Site. There are many others, from cave paintings dating back to the Stone Age to an enormous tree in the centre of the town of Kabwe. These are perhaps less dramatic than the Falls, but have enormous potential for tourism. The National Monuments Commission actually comes under the Ministry of Tourism – all sites are owned by the State – and Katanekwa believes that it is sitting on a potential tourist goldmine.

Conservation is a major part of the Commission's work – declaring new sites before they are threatened by development – and managing and protecting existing ones. Before a Monument is declared, the conservation requirements and cost implications are always taken into account. The Commission has only 40 full-time and 40 part-time staff and it is purely this lack of sufficient manpower, transport constraints and limited financial resources, which restrict the number of sites on their list.

Chapter Three
WILDLIFE FOR PEOPLE

Carefully-controlled harvesting of elephants could provide substantial cash benefits for Zambia. But poachers are slaughtering so many – for their own short-term gain – that the animals are disappearing fast. (CH)

To Use or Abuse?

It is inevitable that Zambia will eventually run out of copper. It is a *non-renewable* resource. But wildlife is *renewable*. If properly conserved, it will last forever and continue to provide food, jobs, medicines and foreign exchange long after the copper-mining industry has collapsed.

There are few countries in the world that can afford to conserve wildlife just for the sake of it. The old belief that wildlife should not be tampered with is unrealistic and impractical in these days of so much human hardship and poverty. Conservation has to work for its living – and be seen to be doing so.

In Zambia, perhaps more than in most other countries, the wildlife is particularly rich and could be put to good use in improving the lives of many hundreds of thousands of people. This does not, however, mean that indiscriminate slaughter is the way to go about it. This may provide enormous cash benefits in the short-term, but Zambia's wildlife resources would soon disappear – perhaps faster than its copper – with such a lack of forethought and planning.

Poaching is evidence of what happens when wildlife resources are abused. Poachers are killing so many elephants – for their own short-term gain – that the animals are disappearing fast. This makes elephants as much a finite resource as copper. Yet properly managed, the same populations could provide a sizeable annual harvest and, through sports hunting fees and ivory sales, could be earning Zambia hundreds of thousands of dollars every year.

Such wildlife resources should be regarded as biological *capital*. Then sustainable use of them, through conservation, can be illustrated as using the *interest* – without eating into the capital itself.

Active wildlife management can also be important for ecological reasons, particularly where the animals are confined to small areas. Fifteen years ago, when elephant numbers were high in South Luangwa and other national parks, they were transforming their own habitat by felling the trees. In this case, carefully controlled harvesting could have saved the elephant population from crashing, given the parks a chance to recover and provided substantial cash benefits – but the poachers got in first. A five-year ban on elephant hunting in Zambia ended in 1987.

A similar situation is now developing in the Luangwa hippo population. With growing pressures from human settlements in the game management areas around the Park, large numbers of hippos are confined to small pockets of land and, as a result, are rapidly degrading the habitat. This is adversely affecting the Park and will ultimately lead to a hippo population crash. Conservation in this case calls for quick action: a carefully managed cropping programme, involving local participation, with the meat and other products going directly to the people of the area or sold to provide investment capital for village projects.

Controlled hunting of this kind can provide a *sustainable* yield of cheap protein for Zambians in areas where there are no commercial butcheries or reliable fish supplies. It already does so in many other countries.

Above:
*The LIRDP is very much a people's
project. Resource conservation
integrated with rural development will
lead to benefits for local people.*

Very often, large wild herbivores produce more protein per hectare than
domestic livestock under the same conditions: in some areas of Zambia,
nearly all animal protein consumed is derived from wildlife.

Less obvious – but equally important – is safari hunting. Licensed
hunting is allowed for a select range of game animals. There are strict quotas
for protected species and a range of hunting fees, from weekly permits and
shooting rights (see table) to payments for exporting animal trophies. Since
hunters tend to spend far more on their safaris than other tourists – sometimes tens of thousands of dollars each in a couple of weeks – the income
from these safari operations is quite substantial.

A very important – and exciting – new initiative in Zambia,
stemming from the National Conservation Strategy, is the Luangwa Valley
Integrated Resource Development Project (LIRDP). The aim of the project is
to integrate rural development in Luangwa with resource conservation. A
range of activities are being developed – such as hippo cropping, game meat
and forest products cooperatives, safari hunting and tourism – in order to
provide a self-sustaining future for both the human and wildlife populations.

Until recently, natural resource exploitation in Luangwa's parks and
game management areas – which collectively total more than 50,000 square
kilometres – has been uncoordinated. Consequently, local people have not
benefited as much as they could have done from the rich wildlife resources,
while the resources themselves have become seriously endangered by
poaching and the lack of proper management.

The LIRDP, which is under the Chairmanship of His Excellency The
President, is very much a people's project. Using local management, it is
specifically designed to benefit, and be managed by, local communities in the
area. It is an important first step in achieving a lasting wildlife future for
Luangwa Valley.

Safari Hunting Fees

	Citizens of Zambia (kwacha)	Overseas Safari Clients (US dollars)
Baboon	10.00	25.00
Warthog	30.00	110.00
Crocodile	30.00	120.00
Buffalo	50.00	180.00
Zebra	80.00	240.00
Hippopotamus	90.00	360.00
Kudu	800.00	800.00
Leopard	1,000.00	1,000.00
Lion	1,000.00	1,000.00

(selection taken from *The National Parks and Wildlife (Licences
and Fees) (Amendment) (No. 2) Regulations, 1984*)

Wild foods form important dietary supplements, particularly in times of crop failure; care should therefore be taken to conserve them.

Top left to right:
Mushrooms, wild ginger fruits, game biltong, prickly cucumber, vinkobala caterpillars, local variety of aubergine, baobab fruit, local variety of bean and native yam.

Fishing

Fish are a relatively cheap – and very important – source of protein in the national diet. Annual production from 1972 to date (with the exception of 1978) has exceeded 50,000 tons, realising over 50 million kwacha every year. But there are signs that more fish are being taken than their populations can withstand and the total quantity reaching the markets has recently been declining.

Although land-locked, Zambia has more than 150 different species of fish and some outstanding fishing grounds. About six per cent (45,000 square kilometres) of its surface is under water. There are five natural lakes (Tanganyika, Mweru–Wantipa, Mweru–Luapala, Bangweulu and Lukanga), three rivers with their tributaries (Upper Zambezi, Kafue and Luangwa) and two man–made lakes (Kariba and Itezhi-tezhi).

The fishing industry provides direct employment for about 250,000 people and indirectly employs another 30,000. Apart from this employment generation, it has also opened up areas that would otherwise never have been developed, through the construction of feeder roads which have been followed by schools and clinics.

Several factors have led to the recent declines in fish production, including inadequate knowledge of the potential of various fishing grounds, which has led to localised overfishing – as well as missed opportunities; bad fishing practices, particularly the use of nets with very small mesh sizes; spoilage of fish due to poor handling and preservation; high demand for the water resources from other users, especially the mining and hydroelectric industries; lack of pollution control for industrial discharges; and inadequate capital investment.

Kapenta drying in the sun.

Left:
Although land-locked, Zambia has some outstanding fishing grounds, which harbour more than 150 species of fish. From top left to right: tiger fish, silver barbel, spotted squeaker and red-breasted bream. (Reproduced with the kind permission of the Chief Postal Manager).

Opposite, Top:
The presence of several large wetlands, lakes and major river systems, combined with a wide diversity of species, endows Zambia with a high fishing potential. The main large commercial enterprises are on Lake Tanganyika and here, on Lake Kariba. These are kapenta fishing boats.

Bottom:
Fishing in Zambia realises over 50 million kwacha every year (1980) but, in addition to the cash incentives, small-scale subsistence fishing is also very important to many people.

Nile crocodiles are taken for granted in Zambia because of their obvious abundance, but well-run crocodile farms are considered by many to be a far better way of exploiting them than poorly controlled hunting.

Crocodile Farming

Zambia is one of the best places in Africa for seeing large numbers of crocodiles. Listed as game animals, they are taken for granted in most places because of their obvious abundance. But there is actually little justification for Zambia's complacent attitude towards crocodile hunting.

Nile crocodiles are common animals throughout most of the country. But local extinctions, caused by over-hunting, are increasingly common; and surveys in recent years suggest that the number of large individuals is declining. A second species, the slender-snouted crocodile, is much rarer and now probably restricted to the Lake Mweru region, in the extreme north. One of the main concerns about crocodile hunting is that no distinction is made between these two species. As a result, the slender-snouted crocodile is in serious danger of extinction in Zambia.

There are several motives for crocodile hunting. Economic motives include local demand for their meat and skins and the need to reduce damage to fishing nets. There is also an understandable desire to kill a dangerous wild animal. But the main economic inducement is the supply of their hides to the luxury leather trade.

While the hunting of wild crocodiles for this trade can be very difficult to control, well-run crocodile farms are considered by many to be an acceptable way of exploiting this valuable resource. Certainly, it would be easier for the Government to justify wide-spread protection for Zambia's crocodiles if the economic returns from such farms were significant.

There are already several commercial crocodile farms in Zambia. The first were established at Lake Tanganyika, in 1979, and Lake Kariba, in 1981. As with most similar operations, for several years they relied upon the collection of eggs from the wild, until satisfactory breeding stocks were established. Wild strains are still sometimes introduced to boost the strength of the stocks. In this respect, the Government is taking a fairly responsible attitude. There is a levy on each egg collected and, once established, the farms are obliged to return a proportion of the eggs they hatch artificially to the wild.

Yellow-billed storks feed by walking slowly in shallow water, with their bills held slightly open, while they stir up worms, crustaceans and other prey items with their feet.

Wetland Conservation

Zambia contains some of Africa's most important wetlands. Together, they cover more than six per cent of the country. Consequently, wetland conservation and development has been identified as an area of top priority in the National Conservation Strategy.

At the suggestion of the Department of National Parks and Wildlife, in the Ministry of Lands and Natural Resources, two areas in particular – Kafue Flats and Bangweulu Basin – have recently become the subject of a major conservation and development project. The aim of this project is to promote the conservation of wetlands by demonstrating how their environmentally sound management can significantly contribute to rural development. In this respect, an important aspect will be the sustainable harvesting of Kafue lechwe and black lechwe.

Bangweulu is comparable in extent to Botswana's Okavango Delta. It consists of three major areas – floodplains, swamp and Lake Bangweulu – and contains no fewer than six game management areas and one national park. Kafue Flats is one of the most important and best-studied wetlands on the continent, much of it inside Lochinvar and Blue Lagoon National Parks and an extensive game management area. The two areas are so enormous that, together, they account for over half all Zambia's wetlands. Both are rich in bird life (including the famous shoebill in Bangweulu and commercially important species of ducks and geese); Kafue lechwe and zebra are particularly abundant on the Flats; and black lechwe and sitatunga are abundant at Bangweulu.

But their long-term future is insecure. In particular, the water regime of the Flats is dependent upon the regulation of hydro-dams, while both areas are experiencing increasing human encroachment.

Yet this encroachment, as well as more formal management practices, is tapping only a small fraction of the wetlands' potential.

Until quite recently, this kind of observation was almost unheard of when developing conservation programmes. The establishment of protected areas – and other conservation activities – have notoriously ignored the needs of local people. This has been true all over the world and, understandably, has often led to varying levels of resentment among the communities involved.

But the Kafue Flats and Bangweulu Basin Project, which forms an important part of the WWF/IUCN Wetlands Programme, is very much a people's project. Since it is designed specifically with rural development in mind, close cooperation with local communities is crucial to its success.

Rhino and Elephant Poaching

Poaching is the single most important wildlife conservation problem in Zambia.

In 1975 there were 8,000 black rhinos in South Luangwa National Park; today, there are fewer than 200 left in the whole of the country. Highly sought-after for their horns, which (although of no real use whatsoever) are more valuable than gold, they have been slaughtered by poachers almost to the point of extinction.

Zambia's elephant population is suffering a similar fate. A decade ago, there were about 100,000 of them in the Luangwa Valley alone. But today only 25-30,000 are left.

In the late 1970s, experts estimated that more than 20 elephants and two rhinos were being killed in South Luangwa *every day*; and the killing was probably as bad, if not worse, in other parts of the country. It was evident that the National Parks and Wildlife Service could no longer contain the poaching problem on its own. Because of the prevailing depressed economic situation the Service was poorly staffed and had very few funds and no equipment with which to fight back.

An independent, charitable organisation – known as the Save the Rhino Trust (SRT) – was therefore set up in 1980, with financial assistance from WWF. With the Government providing rangers and scouts, and all equipment and other expenses supplied with money obtained from a variety of donations and grants from around the world, SRT was in a far better position to combat the poaching problem.

In the beginning, everyone associated with SRT was cautiously optimistic. They believed that the new laws and stiffer penalties, introduced by the Government to back up their efforts, would act as effective deterrents. But their optimism has since waned.

Caleb Nkonga, SRT's Wildlife Warden on secondment from the National Parks and Wildlife Service, takes up the story:

"We are faced with a colossal and daunting task: to patrol some 10,000 square kilometres of Zambia's wilderness. The terrain is bad and my scouts have a really hard time. They patrol on foot, for an average of twenty days every month, searching known entry points, lying in ambush, following trails of destruction left behind by the poachers or using their bush skills to track the gangs down.

"Armed only with hunting rifles – which cannot (by law) be fired except in self-defence – they are expected to tackle determined and dangerous gangs of professional poachers armed with a range of sophisticated automatic weapons. Many poachers use Kalashnikov AK47s, which were left over from the days when Zambia hosted the guerrillas fighting in Rhodesia. The scouts have no legal rights in shoot-outs with poachers – which occur regularly – yet there may be only five of them trying to arrest a gang of more than thirty. Perhaps not surprisingly, several of my men have been seriously injured during their patrols. They really need to be trained in new anti-terrorist skills and tactics.

"For all this, the scouts earn a basic US$30 per month. They have to be very dedicated men – many of the poachers are probably earning six times as much.

"But despite all their problems, the scouts put in an admirable effort and SRT is certainly making headway. Since we began operations about eight years ago, we have made about 1,500 arrests, impounded nearly 500 firearms and recovered 58 rhino horns and 1,260 tusks. The poachers are clearly finding life more difficult with us around; in South Luangwa, they have been forced to work mostly on a hit-and-run basis and now prefer to remain continually mobile rather than to set up semi-permanent camps.

"Nevertheless, there is no doubt that we have so far failed in our task. The rhino population in Zambia is now so low that the poachers operate principally for ivory and regard rhino horn simply as a welcome but incidental bonus. The odds are still heavily on the side of the poacher and we need to be much more effective in the next few years, or we will be left with literally nothing to save."

Caleb Nkonga stresses that SRT's most urgent requirements are more

men and more equipment. There are currently 73 scouts – plus a long-established back-up network of Honorary Wildlife Rangers – working as four separate field units. Each unit is broken down into sections of five, accompanied on patrol by a small team of carriers who transport their rations and any trophies recovered. Since SRT does not have enough vehicles for use on anything other than a collect-and-deliver basis, the scouts have to patrol on foot.

Most sections do not have a radio and, once in the field, are unable to contact their bases again until they rendezvous with the Landrover a week or more later. Nor does SRT possess an aircraft, which is really essential for patrolling much larger areas than is currently possible.

There are several other urgent requirements. The Government needs

to institute far more severe penalties for the poachers. Poaching is at least as great a theft from the nation as emerald smuggling, and as immoral as mandrax drug trafficking. The State has been prosecuting even the highest-placed people for these offences, but not for poaching or ivory and rhino horn smuggling.

More efforts are also required at the business end of Zambia's poaching operations. They are almost certainly controlled by one or two 'Mr. Bigs', as well as an assortment of middle-men, who are well placed to smuggle their trophies out of Zambia and onto the world market. An important international agreement – the Convention on International Trade in Endangered Species of Wild Flora and Fauna (CITES) – to which Zambia became a signatory in 1980, can help the Zambian Police and Special Investigations Teams in this respect.

Most of the horn and ivory ends up outside Africa. If there was no demand, however – because of stricter penalties or changes in traditional beliefs – the poachers would be forced to stop. Rhino horn, for example, is sold in many countries in the Far East, as a fever-reducing cure. Yet scientific evidence overwhelmingly demonstrates that it has no such properties. It is also highly prized for making dagger handles in North Yemen – yet other materials (even other animal horns) could be used in its place.

It may be that the only hope is to concentrate all SRT's efforts in one small area, which still has viable concentrations of rhinos and elephants and where adequate protection can be given round-the-clock. Perhaps rhinos, in particular, should be captured and translocated to this secure area from elsewhere in the country. The day that happens will be a sad one – but it may not be far off.

Save the Rhino Trust scouts and carriers cross a dried-up riverbed at the beginning of a ten-day patrol in South Luangwa National Park. Without a radio – and therefore unable to contact base for help – they may have to tackle gangs of up to thirty poachers armed with sophisticated automatic weapons.

Opposite, Top:
Africa's black rhino population is being decimated by poachers. In 1970, there were 65,000; today, fewer than 4,500 are left. Zambia's rhino population has been hit particularly hard. (CH)

Middle:
Poaching of ivory and rhino horn is losing Zambia millions of kwacha every year, as well as contributing to massive decreases in wildlife populations. Principal Wildlife Scout John Ebanda is seen here standing in the headquarters of the South Luangwa anti-poaching patrol unit; this room contains the skulls of 205 rhinos killed by poachers – representing more animals than the total rhino population surviving in Zambia.

Bottom:
Anti-poaching patrol scouts are armed only with hunting rifles, which cannot be fired except in self-defence. In the future, they may need to be trained in modern anti-terrorist skills and tactics in order to cope with ruthless gangs of poachers.

'The Healer with a Difference': Dr P.J. Mwanza uses wildlife products — which he collects himself — to cure VD, bad spirits, diarrhoea, bilharzia, heart palpitations, nose bleeds, headaches and even madness.

Medicine Man

Once a week, Dr P.J. Mwanza travels into the heart of Zambia's Eastern Province to search for caterpillars, roots, droppings, tortoises, porcupine quills and a variety of other bits and pieces.

Mwanza has been a traditional healer, operating from a stall in Lusaka's Luburma Market, since 1966. He claims to be able to diagnose and treat anything — from bad spirits to diarrhoea — and uses wildlife products exclusively to make his medicines. The roots from various kinds of *Acacia* are particularly valuable; they can apparently be used to treat snake bite, diarrhoea, gonorrhoea and many other complaints.

Traditional healers such as Mwanza were ridiculed by the colonialists. They learned their trade from the spirits — still do — were soon dubbed 'witch-doctors', and became widely considered as barbaric and primitive.

It is therefore ironic that Western medicine today thrives on extracts from many of the same wild plants and animals that Mwanza and his colleagues have been using for centuries. Each year nearly half all prescriptions in the United States contain a drug based on a wild plant, animal or bacterium. Many of the others use substances which were first discovered in wildlife products but can now be synthesised in the laboratory. It is highly likely that

many other natural products – which are used in traditional medicines but often dismissed in the West as 'primitive superstition' – hold the keys to unsolved medical problems.

Mwanza, who is also Provincial Secretary of the Traditional Health Practitioner's Association in Lusaka, maintains that he uses the same basic components as modern physicians – but in different ways. "Grubbing about in miombo woodland, looking for suitable herbal materials, is not far removed from a white-coated chemist grubbing about in his laboratory", observes one of his colleagues.

Many rural people in Zambia – as in many other countries – have no choice but to consult traditional healers. They do not have access to basic health services such as emergency first aid, immunizations and assistance for mothers during pregnancy or childbirth. New hospitals, health centres and clinics are being built all the time but there will still only be 250 trained doctors in Zambia by the end of the decade. The health needs of the people are therefore much greater than modern services can cope with alone.

A large proportion of the population actually *prefers* to consult traditional health practitioners, who usually live and practice among the communities and therefore understand the cultures, beliefs and customs of their patients. Even in Lusaka, where modern health services are more widely available than in most rural areas, Mwanza has more than 400 regular customers – and his list is growing all the time.

Medicinal stall in Luburma Market, Lusaka, displaying tortoiseshell, bird feathers, snakeskin, assorted vertebrae, porcupine quills, honeycombs, lion skin, twigs, live caterpillars, roots, bark and many other traditional medicaments.

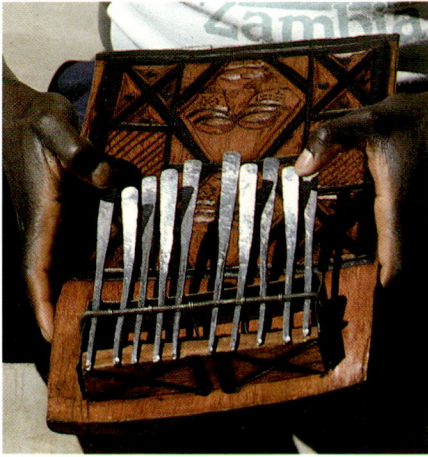

The raw materials used for Zambian handicrafts are largely obtained from wild, uncultivated plants. From top left to right: thumb piano, bark cloth, sisal mat, lupu, bark fibre rope, and beer strainer.

Zambian Handicrafts

Most traditional Zambian handicrafts are strictly practical. Undecorated pottery, animal traps, sleeping mats, spears, tools and food bowls are all produced, in small quantities, for regular daily use. But others are made more for their aesthetic appeal, among them beadwork, carvings, basketware, and the brightly-coloured cotton batik chitenges, which are worn wrapped loosely around the body as the national dress.

The raw materials used are largely obtained from wild, uncultivated plants and tend to reflect the environments in which the handicrafts are made. A basket-weaver could use anything from bamboo, sisal or liana vines to papyrus palms, rushes or grass, depending

For many families, crafts are an important source of extra income and often their only means of earning cash. This is Zintu Handicrafts, in Lusaka, an important outlet for traditional crafts from all over the country.

on where he or she is living. The different-coloured dyes are obtained from various soils, roots, bark and leaves. It is important that agricultural development takes the 'wild' areas, where these raw materials are found, into account much more than it has done in the past.

There is a notable division of labour between the sexes. Both men and women practice basketry – among them are some of the finest basket-weavers in Africa – but, in most cases, the men work in metal and wood while the women tend to specialise in pottery.

Although the traditional skills have not yet been lost, craft-making in Zambia has changed considerably in recent years. Foreign demand has forced people to adapt to suit modern markets since, for many families, crafts are an important source of extra income and often their only means of earning cash. But the change has not entirely been for the worse: education and travel have broadened their ideas.

Chapter Four

THE ZAMBIAN EXPERIENCE

*F*ollowing in Livingstone's Footsteps

Unlike many other safari destinations, where tourists sometimes outnumber vultures around a kill, in the wilds of Zambia you seldom see another soul. The game parks are not well-manicured, with tarmac roads and signposts, but are full of the atmosphere of untamed Africa.

Yet tourism is already Zambia's third largest foreign exchange-earning industry and is rapidly assuming a very important role in its development. Tourists come from as far afield as North America, Europe and the Far East: to see Victoria Falls, one of the most awesome sights in Africa; to participate in the famous walking safaris; to observe, photograph and hunt the wildlife; to fish in some of the best tiger fishing grounds on the continent; to swim and boat at the beach resorts of Kasaba, Nkamba and

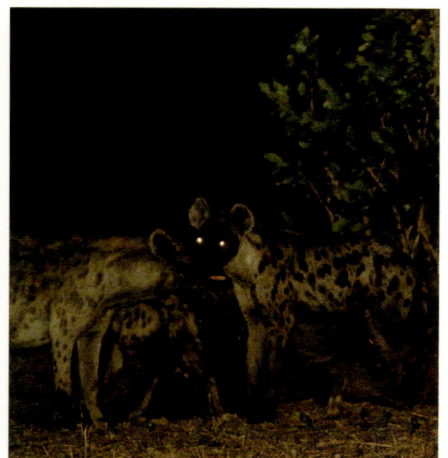

Zambia specialises in open-topped safari vehicles, and is renowned for its night-time game viewing. The spotlight picks out an exciting array of wildlife – including leopards, eagle owls and hyenas – seldom seen by day.

Opposite, Top:
Zambia's national parks are full of the atmosphere of untamed Africa. This impressive herd of buffalo was photographed in Kafue. (CH)

Bottom:
Zambia's wilderness captures the enchantment of Africa as it once was, harbouring a world as remote from the 20th century as almost anywhere you could hope to find on this ancient continent.

Iain Macdonald, Manager of Chibembe Camp, a sprawl of comfortable wooden chalets with thatched roofs on a magnificent bend in the Luangwa river.

Top and Middle:
With some careful stalking, a little patience, and plenty of luck, it is possible to approach a number of Luangwa's animals – in these examples, hunting dogs and an impala – surprisingly closely. (SB, DR)

Ndole Bays; and even to take adventurous whitewater rafting trips down the Zambezi river.

But while Zambia is increasingly becoming aware of its potential as a holiday resort – and desperately needs the hard cash which tourists can provide – the government has adopted a pragmatic approach to the growth of its tourist industry.

Tourism is the fastest-growing foreign exchange-earner in Africa. Many countries have taken advantage of this – and, in the last decade or so, have rapidly developed their tourism potential – but now find that over-commercialisation is taking its toll. The tourist industry in Africa is sustainable only if its unique wildlife, landscapes and traditional cultures – on which tourism depends – are conserved and managed judiciously.

Zambia, however, is a rather late developer in the tourism business. It is therefore in the happy position of being able to learn from the mistakes of these other countries – and is determined to do so. Consequently, the Ministry of Tourism, the Zambia National Tourist Board and the National Hotels Development Corporation are all concentrating their efforts on quality rather than quantity. Zambia, they have wisely decided, is not for the mass market, but for Africa connoisseurs.

The number of tourists eager to experience Zambia's special brand of enchantment is, nevertheless, rising each year. It rose from 56,000 in 1976 to 129,000 in 1984.

Facilities and accommodation are continually being upgraded, to keep pace. But, for the most part, this is being done with due care and consideration. Camps and lodges in South Luangwa National Park, for example, have a friendly and relaxed atmosphere, are comfortable, serve first-rate meals but, most of all, are fairly small and unobtrusive. The authorities, meanwhile, continue to monitor developments carefully.

A Walk on the Wild Side

Walking safaris into the African bush were pioneered in Zambia. The brain-child of wildlife expert and conservationist Norman Carr, at first they were heralded by their critics as dangerous and irresponsible. But twenty years on, having been imitated by several other African countries, they have been dubbed 'the world's most authentic safari' and are more popular than ever.

Iain Macdonald, Manager of Chibembe Camp in South Luangwa National Park, has been a leader on the 'Wilderness Trails' for nearly fifteen years. Born in Northern Rhodesia in 1946, at the age of 28 he was apprenticed to Norman Carr (who then managed Chibembe himself) and has worked in the Park ever since. Iain maintains that walking safaris are the best way to show people from outside Africa what the African bush is all about. Unlike the more familiar pampered outings in zebra-striped mini-buses of some other countries, they offer an exciting sense of danger – unscreened by glass – and give the full impact of the scents and sounds that really bring the bush alive.

The Wilderness Trails, however, are surprisingly free of danger and the critics of the mid-1960s have had to eat their words. Faint-hearted city rats are safe in the hands of armed guards – on loan from the National Parks and Wildlife Service – and experienced guides. The pace is easy but steady, with plenty of time to observe and photograph. It is possible to see a couple of dozen mammal species and two or three times as many birds in a single day.

The clients – never more than six or seven – walk in single file, following instructions from the trail leader. The guard always leads the way. His job is to stand his ground in the face of danger and, if necessary, fire warning shots while the trail leader takes care of the clients. Despite occasional close encounters with lions, elephants, buffalos and hippos, there have been very few accidents. In practice, *shouting* is usually a sufficient deterrent and, on safaris from Chibembe, warning shots have been fired on only three occasions.

Accommodation (for safaris lasting longer than a day) is rough-and-ready, but tremendously atmospheric, in grass huts tucked away in some of the most spectacular corners of the bush. The food is excellent and, while walking, a tea boy always brings up the rear, balancing a kettle, mugs, fresh water, milk, sugar, coffee and tea on his head.

Chapter Five

FORESTS FOR THE FUTURE

Trees mean many things

Most of Zambia is covered with forests and woodlands of one kind or another. Miombo woodland – an open mixture of shrubs, various trees of moderate height and tall grasses – covers about 70 per cent of the country; several other varieties also exist, their type and extent depending largely on altitude and rainfall.

      This vast resource is capable of providing a wide range of benefits for the people of Zambia. Timber provides construction material for houses and other buildings; the railway lines run on wooden sleepers and the Copperbelt mines are literally supported by wooden props. Wood is used to make furniture, tools, telegraph poles, fences and many other commodities; it is essential for cooking and heating in nine out of every ten Zambian homes;

Miombo woodland – an open mixture of shrubs, various trees of moderate height and tall grasses – once covered about 80 per cent of the country. The climax vegetation remains miombo – but in many areas the trees themselves have been removed. (MC)

Charcoal is used in 24 per cent of rural households and 87 per cent of urban households. There are very few alternatives; only a few per cent of houses in Lusaka, for example, are serviced with electricity.

and its pulp is used for making paper. There are also innumerable employment benefits from the many forestry-related jobs associated with these products.

Whole forests provide protection for the soil, by shading it from the drying effect of the sun and breaking the destructive force of heavy rain; and they ensure a reliable water supply, by acting as vast 'sponges' which hold water and release it gradually. Properly conserved forests also provide homes for wildlife, which in turn can provide food, fodder, medicines and the basis for a major tourist industry.

If Zambia were to utilise its forests and forest products sustainably, in addition to all these benefits it could save foreign exchange by reducing a wide range of forest product imports; foreign exchange could actually be created by exporting many of these items.

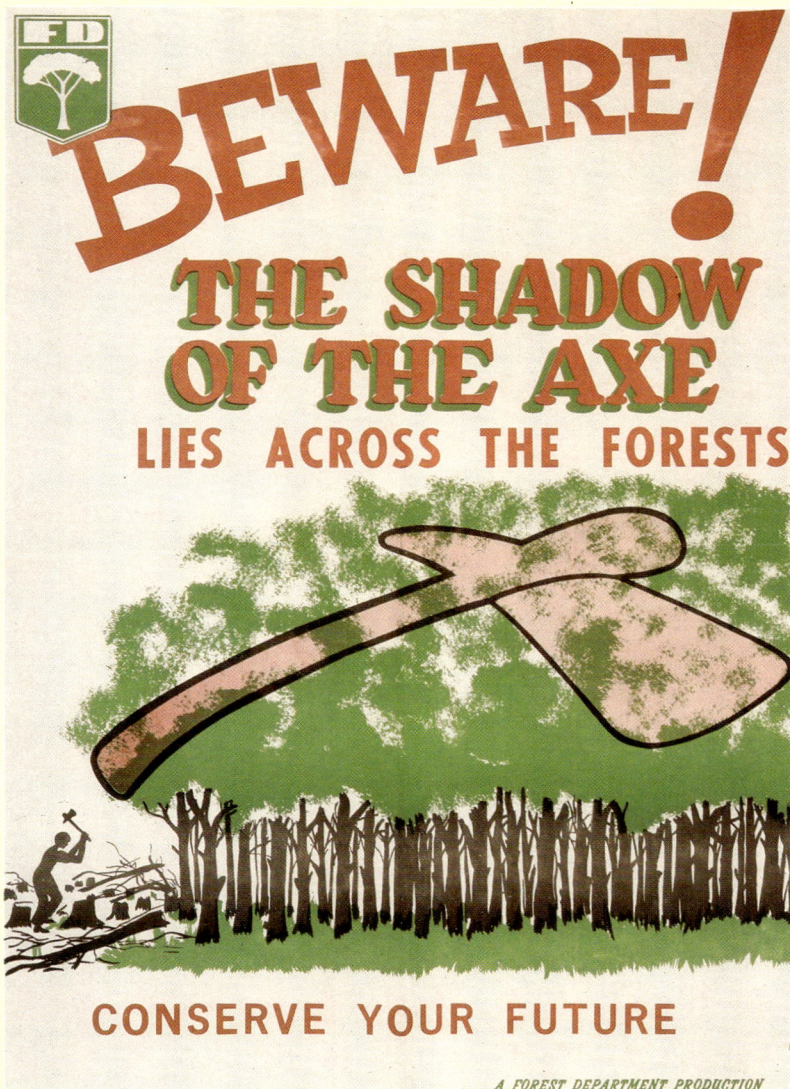

Forest Department poster.

The Vanishing Forests

Deforestation is not *yet* an overwhelming problem in Zambia. But while most of the country's natural woodland is not under immediate threat, the indiscriminate cutting of trees around many towns and cities is a major cause for concern.

Zambia loses about 0.5 per cent of its woodlands every year; around large human settlements the loss is considerably higher. Most of this deforestation has gone ahead randomly, unplanned and often for short-term profits rather than to cater for real local needs. There are six main causes:

Cutting of firewood and charcoal
This is the major cause of deforestation around the larger towns. Traditionally, the source of fuelwood is the dry and decaying branches or felled trees from

land cleared for agriculture. But the increased felling or lopping of live trees in other areas signifies a growing shortage – to such an extent that fuelwood has entered the market place – and its inflated price has become a heavy burden for poorer town-dwellers. The problem is worst in those areas supplying fuelwood to the growing urban populations of Lusaka and the Copperbelt towns.

But fuelwood cutting does not usually amount to complete clearance – since certain species of tree are preferred to others – so the problem often appears to be less dramatic than it really is. In addition, women (who are the normal fuelwood collectors and have to travel further and further afield to find enough wood) are usually less vocal in such community matters. The combined result is a general lack of awareness – and, consequently, planting trees for fuelwood use is practically unknown.

Chitemene
This is a form of shifting cultivation, involving the lopping, and sometimes felling, of indigenous trees and the burning of the cut wood to generate mineral ash for fertilising the soil. The resulting gardens are cultivated for about six years, usually under finger-millet, and are then abandoned for a further area of new woodland. This works well for low population densities – because the soil is given sufficient time to recover before being re-used – but is quite unsuitable for a population density of more than four people per square kilometre. At least 10,000 square kilometres of natural woodland are cut every year by farmers who practise chitemene.

Clearing land for large-scale agriculture
Clearing land of its forests, to make room for settled agriculture, is a major cause of woodland loss on the national scale. In many areas this has exacerbated the shortage of fuelwood and other forest products, since the trees are often burned to waste rather than put to good use.

Overgrazing
Although heavy elephant pressure in areas such as South Luangwa National

The chitemene system of shifting cultivation. This works well when the human population density is low but, as the demand for land increases, it is becoming a serious cause of woodland destruction. (SMJB)

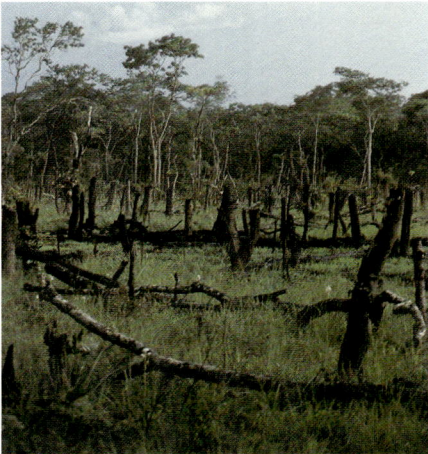

Clearing land of its forests to make room for settled agriculture is a major cause of woodland loss in Zambia.

Bush fires can be catastrophic to natural woodland, especially if they occur late in the dry season, when over 75 per cent of all trees under three metres high may be killed.

Park has destroyed areas of mature woodland, overgrazing very rarely leads to the removal of mature forest. Its most serious effect is the inhibition of regeneration *after* fuelwood cutting, chitemene or land clearance.

Forest fires
Fires are potentially the greatest enemy of Zambia's forests. Many, such as those caused by discarded cigarettes or camp fires, are the result of carelessness. Others are started deliberately, perhaps to drive animals into traps or for robbing bees' nests. Most serious of all is the late season burning (from September to the beginning of the rains in November) that is sometimes carried out to maintain pastures on ranches or in game management areas, as well as for chitemene. Fires at this time of year, when the vegetation is dry, can spread easily. They completely destroy regenerating seedlings and often kill mature trees and many woodland animals. They sometimes even destroy crops and villages and kill people.

Overcutting
This is the result of both legal and illegal, large- and small-scale, operations removing trees wholesale for construction, without leaving adequate young stock or the right conditions for regeneration. A particularly large impact on forest resources is made in this way by commercial sawmilling.

The consequences of deforestation are serious and involve a complex web of environmental problems. If allowed to continue unchecked, these could have disastrous consequences. Large areas of land would turn into desert, there would be serious landslides and floods, rivers and streams would dry up, dams would be blocked by sedimentation, wildlife would disappear and there would be a critical shortage of important forest products.

A widespread lack of fuelwood would then cause the nutrition of the population to deteriorate, as less and less is used for preparing food; there would also be a decline in productivity as more human effort and time have to be invested in procuring fuelwood; and there would be an inevitable reduction in outlay on other basic needs, as more money is diverted to energy acquisition.

Signs of some of these environmental problems are already appearing in some areas.

Zambia's forests are capable of providing a wide range of benefits for people. But they are being abused and, without careful management, will not continue to provide unlimited resources for much longer. (MC)

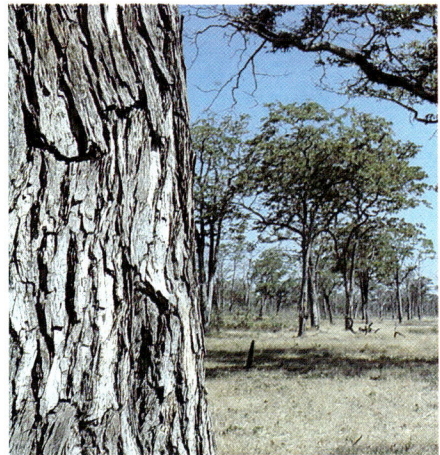

Forest Management

There are many ways of reducing the problem of deforestation. The establishment of carefully managed forest reserves, education and awareness campaigns, legislation and its enforcement, land-use planning, the provision of more efficient ways of using forest resources and improved plantation forestry all have important roles to play.

Nearly 70,000 square kilometres, or 9 per cent, of Zambia's land area is under some form of forest management. Classified as either *protection* or *production* reserves, they are all owned and managed by the State. The main function of production reserves (which account for two-thirds of these forests) is to produce wood on a sustainable yield basis. Protection reserves are designed to maintain freshwater supplies and prevent soil erosion; exploitation of these is either entirely prohibited or limited in such a way as to improve the forest stand and growth. The Government has set a target to increase the total area of these State-managed forests to at least 15 per cent.

Present education efforts, by the Forest Department and the Natural Resources Department, are admirable but pitifully small in relation to the scale of the problem and its highly dispersed nature. Similarly, while the Forest Act makes the right provisions, enforcement by the Forest Department is necessarily limited to controlling commercial exploitation within reserve forests; stumpages are too low and fees collected probably represent less than ten per cent of the wood actually cut.

Land use planning – in other words, allocation of land according to the true capability of its natural resources – is not carried out. Besides, it would be difficult to implement unless there were up-to-date inventories of

wood stocks and growth rates. More work is also required on efficient wood stoves and charcoal kilns. They are being developed but have yet to be put into general use.

Reforestation is needed to counter the problems of deforestation; more especially, fuelwood plantations are needed to cater specifically for the growing fuelwood demand. But the problem with current plantation forestry is that the right kinds of forest do not exist in the right places. There is great potential, for example, for planting tree crops on farmland with poor or eroding soils. These could make use of non-arable land, to produce fruit, nuts and honey (in areas where tilling the soil would dry it or make it susceptible to erosion) and to provide shade, shelter, soil nitrogen and fuelwood, as well as foliage browse for livestock.

The best overall strategy for forest conservation in Zambia is community and farm forestry, through a variety of self-help schemes. Wisely run, these could solve the problems of labour and management costs, delayed financial return, distance from source to demand, law enforcement, public understanding of the value of forests and the increasing costs of fuelwood. The active support of the people, through this kind of approach, is undoubtedly essential if Zambia's forest resources are to be successfully conserved and developed.

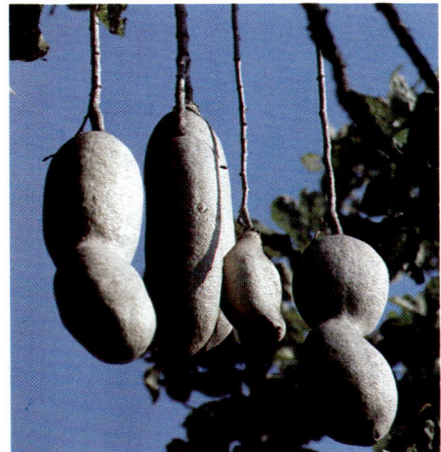

Left:
Baobabs have many uses: the fruit pulp makes a refreshing drink (its vitamin C content is one of the highest known in a natural fruit); the young leaves can be used like spinach, or to make soup; the wood is used for fishing floats; the inner bark yields a strong fibre for rope-making; the seed pods are used for carrying liquids; and the hollow trunk was once even used as a prison cell. (MC)

Below:
Each hanging fruit of the sausage tree may be over half a metre long and weigh up to four kilograms. Although poisonous to humans when unripe (and inedible when ripe) they are used to ferment beer and as a source of red dye. They are eaten by squirrels, mongooses, baboons, hippos and rhinos. (SB)

The Tree's Prayer

Ye who would pass by and raise your hand against me,
Hearken before you harm me.
I am the heat of your hearth on cold winter nights,
The friendly shade screening you from the summer sun.
And my fruits are refreshing draughts,
Quenching your thirst as you journey on.
I am the beam that holds your house,
The board of your table,
The bed on which you lie,
And the timber that builds your boat.
I am the handle of your hoe,
And the door of your homestead,
The wood of your cradle,
And the shell of your coffin.
I am the gift of God,
And the friend of man.
Ye who pass by, listen to my prayer
And harm me not.

(Published by the Forest Department)

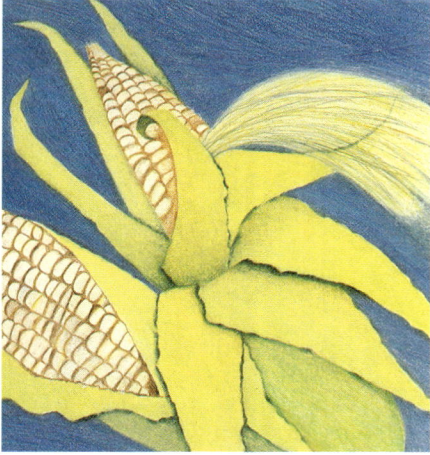

Chapter Six
FOOD FOR THE NATION

A Country of Farmers

Opposite:
Millet is more suitable than maize in drought-susceptible areas, but personal preference and the unavailability of seeds limit its use in many areas.

Below:
Levels of mechanisation remain low in Zambia, and the use of animal draught power is only just developing. Machinery such as this is only found on the largest commercial farms. (IM)

Almost one-fifth of Zambia is under cultivation and seven out of every ten working people are engaged in farming in one way or another.

Zambian agriculture is characterised by a distinct contrast between commercial and subsistence farming. Large-scale commercial operations are concentrated around the central railway line. Subsistence farms – more than 600,000 of them – are distributed throughout the rest of the country's arable regions.

Maize is the most important food and cash crop, followed by sorghum and cassava. But farming in Zambia is mainly rain-fed (only one per cent of the potential agricultural land is irrigated) and recent droughts have shown that maize is not always reliable. Farmers in drier areas are therefore being

Large-scale farming operations near Lusaka. Zambian agriculture is characterised by a distinct contrast between commercial farming, such as this, and subsistence farming. (MC)

encouraged to concentrate more on sorghum, millet and even certain wild plants, which are all much more expedient. But many factors, including prejudice, personal preference and a lack of suitable seeds, limit the possibilities for such a transformation.

In the villages, agricultural work is done mostly by the women, while the men (who may have several wives) stay at home. The level of mechanisation remains low and the use of animal draught power is only just developing.

Since Zambia can no longer trust in copper – and agriculture is the major alternative for earning foreign exchange – it needs to export sufficient cash crops to supplement the declining income provided by the mining industry. But production of almost all crops falls below the required level – and much food still has to be imported. In an effort to make Zambia self-sufficient in the future, the Government has launched a nationwide agricultural programme, aimed particularly at smallholder cultivators.

Maize is Zambia's most important food and cash crop, though recent droughts have shown that it is not always reliable.

Opposite:
Zambia will soon be self-sufficient in tea, which is becoming an increasingly important crop. (IM)

Cattle are the main source of wealth for many people in Southern, Central and Eastern Provinces. The number of traditionally managed cattle at the end of 1983 was 2,048,000.

Chickens for sale in Lusaka's Luburma Market.

Goats can be a good source of income provided they are well managed. But in some areas they have been allowed to eat almost everything, thus turning the land into a series of dust bowls.

Livestock

Livestock numbers in Zambia are so high that they are causing severe environmental problems. Overgrazing results in soil erosion, while selective grazing of the tender, more palatable and productive grass species encourages the spread of coarser ones. But action to reduce these problems – through better rangeland management – has been limited on the Government's part, and is almost non-existent on the part of the farmer.

Ninety per cent of the cattle in Zambia are owned by traditional herders. Important as status symbols and sources of wealth, they are not killed for meat and are rarely sold. Only about five per cent are marketed each year. Their main practical uses are for dairy products and as a source of manure and animal power in other farming activities.

Herding is usually on a free-range system, with a total of about 10 million hectares of grazing land throughout the country. But only one quarter of this land is available for dry season grazing, which results in overcrowding in dambos and on floodplains for a major part of the year. Yet no special provision is made for dry season grazing: a village's pasture consists merely of the part of the village lands not under cultivation.

Poor pasture, disease and, in particular, a lack of management make Zambia's livestock unproductive and unmarketable. However, with better management of livestock and pasture, productivity could be improved considerably. And when more animals become marketable, they could be removed from the rangeland, so that farm income is increased and over-grazing is reduced.

The problem of rangeland degradation is deeply rooted in traditional values and in poorly planned rural development. It is also a legacy of the colonial era. Africans and their livestock were forced into 'reserves' or 'trust lands' on poor soils. The increased livestock densities and the erodible soils inevitably led to overgrazing damage.

But while the problem clearly cannot be solved in a few years, there are a number of steps which can be taken to improve the situation. A current programme of the Department of Agriculture, for example, is to establish Cattle Development Areas where the local population is encouraged to adopt modern animal husbandry methods. These include active health care, and pasture management and improvement. The aim – to increase animal offtake – fulfills many conservation objectives as well as those of development.

Tsetse Control

Approximately one-third of Zambia is infested with tsetse fly. The vector of sleeping sickness, which affects both people and cattle, it has been the subject of extensive eradication programmes in the last thirty years.

Before insecticide use became widespread, the main tool of control was bush clearance and game elimination. But this was expensive and inefficient. In the absence of agriculture, the bush returned – and so did the fly. Aerial spraying with low doses of endosulphan is the most common method used today. Ground spraying of the (more dangerous) persistent organochlorines, DDT and Dieldrin, is limited to smaller areas or places with more difficult terrain.

Increased tsetse control is being considered as part of a regional project to remove the fly belt common to Zambia, Zimbabwe, Malawi and Mozambique. The economic benefits of this will depend very much on whether the livestock owners will be able to increase significantly their productivity in tsetse-free areas.

The effects of tsetse control measures on non-target organisms, and in terms of soil and water pollution, is probably bad, but relatively unknown. More environmentally sound techniques – such as insecticide-impregnated screens and traps, or fly sterility programmes – are currently being investigated. If any of these control measures is effective, there is no doubt that the ensuing increased cattle population could lead to overgrazing and soil erosion; it is therefore essential that tsetse control goes hand–in–hand with careful livestock development and land use planning.

Agricultural Chemicals

Zambia spends three million kwacha every year on importing agricultural pesticides. They are the main weapons against pre- and post-harvest crop losses, which in most developing countries are in the region of 30 per cent. Their use in weed control also makes labour more efficient and, by preventing massive periodic losses (caused by insects and diseases) they stabilise agricultural production. Cotton would be virtually impossible to produce without pesticides, and yields of other crops would be much lower.

Zambia clearly gains a great deal of benefit from these chemicals. But control of their use – the type of chemical, its method of application and the quantity used – is almost non-existent. Containers are rarely labelled to warn of their lethal contents; drums leak; and workers applying the chemicals do not wear protective clothing. As a result, Zambia has been victimised by certain chemical companies (and some aid agencies) and is being sold banned, below standard or old pesticide products.

The threat is particularly severe because the chemicals most commonly used in Zambia are extremely dangerous. They are mostly organochlorines – in particular, DDT, Dieldrin, Aldrin, BHC and Heptachlor – which have been banned in many other countries. Organochlorines take years to decompose and, as they become concentrated in the food chain, can kill a wide range of different animals.

Unnecessary – but routine – spraying is also causing environmental problems. The pesticides build up in the soil and, instead of killing target species attacking crop plants, are destroying useful soil organisms. In other cases, where the pesticides are used year after year, resistant strains of pests are developing. Some 300 species are known to have developed resistance to commonly used pesticides in this way.

*S*oil Erosion and Infertility

Soil takes hundreds or even thousands of years to form but can be destroyed or lost in a matter of hours. There is strong evidence that Zambia's cultivated lands lose more than three million tonnes of topsoil every year. Soil erosion has become one of the most serious environmental problems facing the country today.

Ultimately, all life depends upon the thin layer of soil on the surface of the earth. If knowledge about its conservation is lacking, or is abandoned in favour of methods producing the highest output as quickly as possible, it becomes prone to erosion. Zambia's soils – with a few notable exceptions – are generally of poor fertility and are particularly susceptible to erosion.

Erosion can be caused by, among other things, deforestation, over-grazing, burning or bad agricultural practices. All these activities remove vegetation cover and expose the soil to the elements. In serious cases, it can be washed away in heavy rains and, as run-off water gains momentum, small channels – and soon great gullies – are formed. The eroded soils rob farmers of their livelihood, and build up in irrigation channels, rivers and reservoirs, reducing their effectiveness and sometimes causing floods.

Certain cultivation techniques can cause more soil problems than others. Ploughing up and down the slope, rather than across, can accelerate erosion; and row crops, or crops such as sunflower which do not give complete ground cover, can encourage erosion if grown from year to year. This sort of problem is widespread; it does not necessarily lead to catastrophic erosion but is certainly debilitating for productivity.

In pre-independence times, a considerable commitment and effort was invested in soil conservation. Conservation works were constructed on arable lands, most of the cultivated fields were contoured, and storm drains, windbreaks, banks or terraces were common. Catchment plans were operated and implemented and good land management advice was given to farmers.

There are still many areas (such as in the Ngoni Reserves of Eastern Province) where the land continues to be well protected with small fields, contour ridging and strip cropping. But a number of things have changed since the colonial days. The economy has become more monetised, hence farmers are responding most to those activities which produce direct economic benefits – often at the expense of conservation. Soil conservation works have fallen into disrepair – due to limitations of labour and finance – and farmers receive scant advice on soil conservation.

Rangeland degradation – caused by both domestic livestock and wildlife removing the vegetation and compacting the surface with their hooves – is another major cause of erosion. It is most critical around drinking places and along tracks near kraals. In Eastern Province and parts of Southern Province, the system of herding cattle in dambos during the rains is causing severe gulley erosion. Affected dambos are no longer able to retain enough water to last through the dry season.

Controlled burning, carried out as a measure to improve grazing (it favours regeneration of grasses rather than woody species) is also an important cause of erosion. If poorly controlled – and especially if it occurs late in the dry season – the great heat of the fire (up to 900 degrees centigrade) destroys the structure of the topsoil and removes all the protective vegetation. This leaves the surface highly vulnerable to erosion in the first rains.

Some of the worst cases of soil erosion are in township areas. The main problem here is road-building, which suffers from incorrect design and layout and inadequate drainage. Roads in some areas have often become impassable due to gulley erosion, posing a severe hindrance to rural development.

Opposite:
Soil erosion can destroy, in just a few hours, a resource which has taken hundreds or even thousands of years to develop. It is a widespread problem in Zambia – millions of tonnes of topsoil are needlessly lost each year from cultivated lands. (MC)

Soil infertility is also thought to be a growing problem in Zambia, caused by continuous planting of a single type of crop. This is causing nutrient deficiencies and imbalances, and hinders seedling emergence and root penetration. Yet continuous cropping of maize is common. However, some heavily fertilised soils are actually experiencing declining yields, because the heavy doses of fertiliser have acidified the soils. Others have become infertile through leaching – due to insufficient organic matter to retain the nutrients – largely as a result of inadequate manuring and overcultivation.

Finally, a problem which is not well understood in Zambia, but which will become increasingly threatening as more irrigation schemes are developed, is soil salination. Salt accumulations are already to be seen in some fields of the Nakambala Sugar Estates. Incorrect drainage is considered to be the main cause. Proper drainage is an essential, but often neglected, element of irrigation schemes and should always be included in order to lower the water table and prevent salt accumulation.

Most crop production in Zambia is rain-fed, but irrigation schemes are being developed in many areas. Adequate drainage is an essential element of these schemes, to avoid problems of soil salination.

Opposite, Top:
With the clapping of thunder and flashes of lightning, the skies finally open in late October or early November. The parched, cracked earth turns into a bog overnight. (CH)

Bottom:
By African standards, Zambia is considered to have an abundant water supply. One of the key conservation and development issues in the country today is how best to use this valuable resource. (MC)

Water

Water is vital to all forms of life. It is used for almost every daily activity, such as drinking, cooking and washing, as well as for industry, hydroelectric power and, of course, agriculture. Yet its importance is realised only when it is not available – without reliable water supplies, all these activities suffer. Every day water is taken for granted and very little care is taken to conserve it.

By African standards, Zambia is considered to have an abundant water supply. It contains, or has access to, several large lakes, major rivers and swamps and generally enjoys better groundwater conditions than most surrounding countries. But clean drinking water is by no means accessible to all; and, since most crop production is rain-fed, the recent drought has proved a major constraint to agricultural progress.

A survey in 1980 indicated that about 40 per cent of Zambia's rural population did not have reasonable access to clean water. Most urban dwellers have access to piped water, but it is often polluted with industrial chemicals and sewage. The cesspits of many parts of Lusaka, for example, lead directly into drinking water aquifers which supply boreholes.

Water is in particularly high demand in agriculture, with the stress now put on development in this sector by the Party and its Government. Better water management practices, which would capture the little rainfall that Zambia receives, are needed urgently. This requires intensive watershed management, which should include forest and grassland protection, various water harvesting techniques and engineering devices such as dams and boreholes. These programmes are costly but essential.

With growing demands on water for agricultural, industrial and domestic use, Zambia could experience a severe water shortage by the year 2000. Current trends also suggest that this shortage might be exacerbated by industrial pollution spoiling the diminishing resources. But with careful management and appropriate environmental planning, a severe shortage of this kind could be avoided.

Chapter Seven
THE DEVELOPMENT CONTEXT

Fuelwood is the principal domestic source of energy in Zambia, providing 80 per cent of the energy used in towns and cities and meeting virtually all energy requirements in rural areas. However, as the tree line around settlements recedes due to overcutting, supply becomes more of a problem.

Energy

There are four main sources of energy in Zambia: hydro-electric power, imported oil, coal and wood. Fuelwood is the principal domestic source, providing 80 per cent of the energy used in towns and cities and meeting virtually all energy requirements in rural areas. As a result, the impact of tree-cutting extends far out from many urban centres – as much as 200 kilometres by road around Lusaka and the Copperbelt towns – and a great deal of time and money has to be spent on obtaining this fundamental resource. A number of pilot projects are currently researching the use of cow dung, grass and agricultural wastes to produce biogas, and the use of wind-powered pumps and solar-powered cookers, as possible alternatives.

Hydro-electric power, on the other hand, is in abundant supply. The total output of all Zambia's power plants – the major ones are Kafue Gorge,

Dams such as Kariba bring benefits to the nation through hydro-electric power production, and act as reservoirs for water which can then be used for fisheries, drinking supplies and agriculture. However, they must be properly planned from the outset, if the long-term environmental cost to society is not to exceed these benefits. (IM)

Kariba North and Victoria Falls – is about 1,738 MW. Even this is more than can be used at present – and it is estimated that the potential capacity is nearer 4,000 MW. In addition, the production cost of hydro-electric power in Zambia is one of the lowest in the world.

But this ample supply of electricity is not made readily available to the public and only a very small proportion of urban households are connected to it. The main constraint to its extension is the prohibitive cost of electrical appliances.

Unfortunately, there are also constraints on the use of electricity in the mining and manufacturing industries, most of which are petroleum-based and not geared to rapid change. Imported oil costs are therefore high – currently about 16 per cent of Zambia's export earnings – especially since

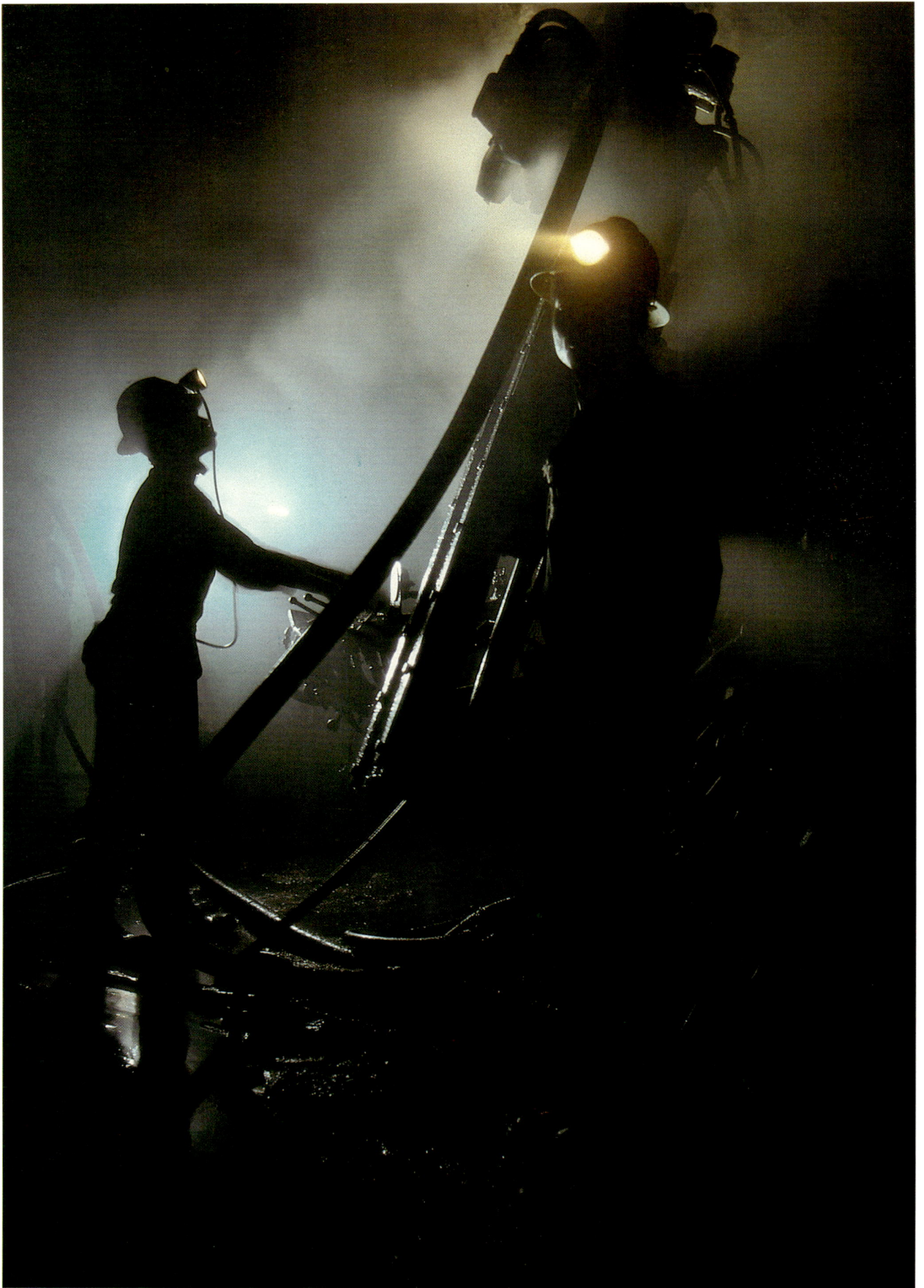

long road distances are involved and the cost of maintaining the oil pipeline to Dar es Salaam is considerable.

However, in developing Zambia's electricity potential, the long-term consequences of building dams for hydro-electric power need to be taken into account. The environmental costs can sometimes outweigh the benefits if they are not planned properly from the beginning – as experience has shown in the past. The Itezhi-tezhi Dam, for example, has made significant impacts on the wildlife of the Kafue Flats and on the Nakambala Sugar Estate – which should have been assessed prior to the dam's design and construction.

Mining and Industry

Zambia's copper mines have been worked for several thousand years. At first, the copper was extracted only from surface deposits. But the advent of mechnanisation in the 1920s gave access to the much richer deposits hidden almost entirely beneath a thick blanket of soil, laterite and rocks. Today, about 600,000 tonnes of copper are produced every year, making Zambia the fifth largest producer in the world.

For half a century, copper exports – to a score of countries on all five continents – have provided more than 90 per cent of Zambia's export earnings.

Copper is used mainly by the electrical industry, for high conductivity wire, as well as in the engineering, transport and construction industries. Much smaller amounts are used in consumer goods, coinage, ammunition and as mineral salts in agriculture. But these days copper is not the prized metal it once was (largely due to the silicon chip and fibre optics) and, in real terms, its price has halved in the two decades since independence.

Although Zambia's copper reserves are rapidly running out, and prices are sliding, the mining industry is still the largest contributor to national income. It is also the largest reservoir of skills and expertise. Its labour force includes over 3,000 different job categories: engineers, metallurgists, nurses, accountants and computer experts among them.

Inevitably, mining has had a profound effect on the history of the Zambian people, pulling them by the thousand from rural areas to the famous 'Copperbelt' region in the north of the country. Covering an area of about 50 kilometres by 110 kilometres, the Copperbelt is the biggest industrial concentration in black Africa. It contains about 6 per cent of the world's proven copper reserves (including a few smaller deposits scattered throughout the country) and five major mines. The seven mining towns in the region – Kitwe, Ndola, Chililabombwe, Chingola, Kalulushi, Mufulira and Luanshya – have a combined population of some 1.5 million.

Zambia also mines cobalt – it ranks second in world production and possesses almost eight per cent of the world's proven reserves – as well as smaller quantities of lead, zinc, coal, emerald, amethyst, tin and other minerals.

Until 1975, the manufacturing industry constituted Zambia's largest non-mining economic sector. But recently most industries have been operating far below their capacities and have become very consumer-oriented. Foodstuffs, beverages and tobacco are the most important, in terms of employment, investment and output, followed by textiles, sawmilling and the manufacture of cement products, chemicals and pharmaceuticals. The greatest growth in manufacturing, however, has been in the newer industries such as paper, rubber and plastic products.

Industrial pollution is not yet a nationwide problem in Zambia. However, it is serious locally where mining and industry exist. The areas most prone to pollution are the Copperbelt, the Kafue Industrial Centre and Lusaka. In particular, mining results in large quantities of waste rock, contaminated water, tailings, slag and various effluents and emissions. These have contributed significantly to water pollution, acid rain and a build-up of toxins in the soil.

Many companies – including Zambia Consolidated Copper Mines Limited – have made admirable efforts to reduce their impact on the environment. But financial constraints mean that not enough is being done and, meanwhile, many other companies are doing nothing at all.

Copper is used mainly by the electrical industry, for high conductivity wire, as well as by the engineering, transport and construction industries. Much smaller amounts are used in consumer and other goods.

Opposite:
Zambia's copper mines have been worked for several thousand years. Today, about 600,000 tonnes of copper are produced every year. But the mining industry's days are numbered. (IM)

Overleaf, Top:
These days, copper is not the prized metal it once was, largely due to the silicon chip and fibre optics. (IM)

Bottom:
Cobalt processing: Zambia ranks second in world cobalt production and possesses almost eight per cent of the world's proven reserves. But recently, with the lack of demand in the industrialised world, cobalt has been sliding in price. (IM)

Chapter Eight

LOOKING
TO THE
FUTURE

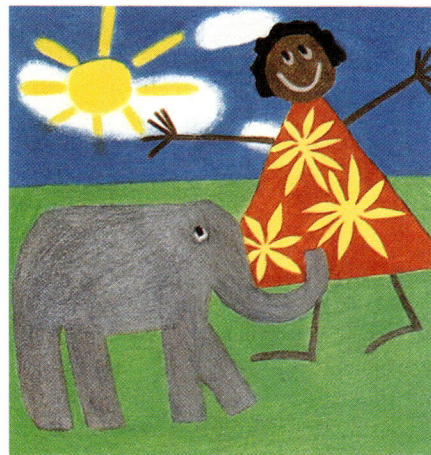

*P*opulation Pressures

At the beginning of this century, Zambia had a population of about one
million people. By the time we enter the next century, it is expected to have
reached ten million. Since each new arrival must be fed, clothed, educated
and housed, the demand for natural resources will have increased at least
tenfold in the space of just one hundred years. But can Zambia's resources
continue to meet this growing demand? If not, the only alternative is to
stabilise the growth of the population.

There is, of course, enormous potential for making better use of
existing natural resources. This is what the National Conservation Strategy is
all about. But Zambia quite clearly cannot provide for *unlimited* numbers of

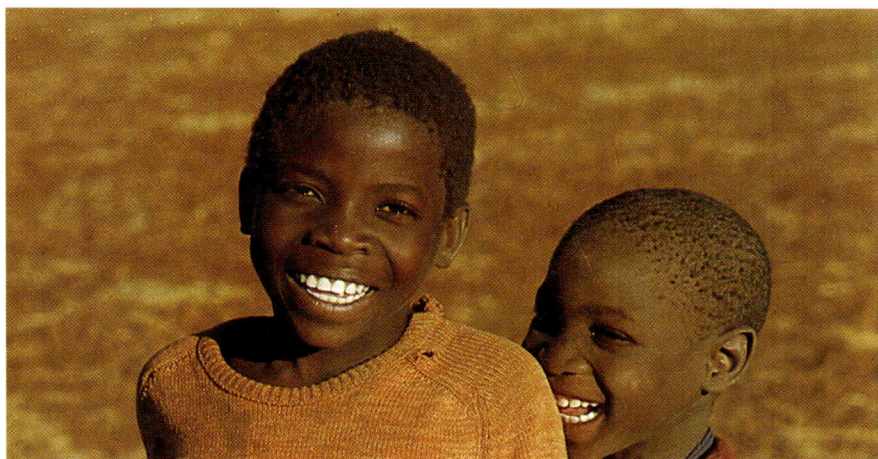

*As we approach the next century,
Zambia's population is expected to
reach ten million. But the country's
natural resources cannot provide for
unlimited numbers of people.*

people. Annual population increases are already severely eroding most
attempts to raise the standard of living.

A national policy for population control has been advocated in
Zambia for several years but does not yet exist. However, there are voluntary
organisations such as the Zambia Planned Parenthood Association which, in
collaboration with the Ministry of Health, try to influence couples to practice
birth control and plan for family growth. These voluntary bodies receive
little direct support from Government, although they have had considerable
moral support.

It is now imperative to take this work a stage further and to develop
a national policy as a matter of priority. Without such a policy, which would
limit the demand for resources, all the positive impacts of national development will be diluted and environmental problems will continue to get worse
rather than better.

Above:
Environmental education must reach adults and children alike.

Right:
Children are, in general, more receptive to new ways of thinking about the environment than adults. (MC)

Caring about Conservation

A comprehensive programme of conservation education is perhaps the best way of ensuring the long-term conservation of Zambia's natural resources. People who appreciate the importance and advantages of conservation are more likely to accept it as a natural part of their daily lives.

Zambia entered independence with only 380,000 primary school pupils. Today it has more than 1.2 million. Secondary school enrolment has also increased, from 14,000 to 115,000, over the same period. But environmental awareness is still very low. In all the local languages, such as Nyanja for example, no distinction is made between *game* and *meat* – *nyama* serves for both; and there is no local word for poacher – *hunter* is used instead, as the concept of poaching is very new.

In recent years, however, substantial progress has been made with conservation education programmes for young people. In particular, two conservation clubs have managed to reach more than one in five of Zambia's school children.

The first to be launched, with support from WWF and Bata, was the Chongololo (or Millipede) Club, in 1973. Despite the erratic delivery of Club magazines and accompanying Teachers' Guides – due to constant shortage of funds – the Club has become a truly national movement for environmentally-minded young people. There are now 1,000 registered groups involving nearly 40,000 children between the ages of seven and fourteen. A weekly thirty-minute radio programme on conservation – called the *Chongololo Club of the Air* – helps to reach an even wider audience.

In 1980, a sister organisation – the Conservation Club – was launched, to cater for young Zambians in secondary schools and colleges. It already has 150 active groups, involving more than 5,000 students.

A special Field Education Unit conducts training courses for Club leaders. These generally last for up to five days and cover wildlife conservation and management, population, forestry, agriculture, fisheries, nutrition, pollution and many other topics involving Zambia's natural resources. In recent years, the work of the Clubs has included wider conservation issues than just wildlife, which was the main focus of attention in the early days.

In the absence of a comprehensive environmental education syllabus in Zambian schools, many pupils and teachers have come to value the Clubs as important extra-curricular activities. This is a sentiment shared by the Ministry of General Education and Culture, which actively encourages the formation of Chongololo and Conservation Clubs in schools.

Children are, in general, more receptive to new ways of thinking about the environment than adults and, of course, are the decision-makers of the future. But the environmental problems of *today* are so acute that Zambia can't just wait for this new generation of enlightened children to grow up. Environmental education must reach young and old alike. The need for conservation education in every sector of adult society – from politicians and industrialists to scientists and subsistence farmers – is therefore paramount.

Conservation is about looking after natural resources so that future generations may also benefit from them.

Right:
Secondary school enrolment has increased eight-fold since independence.

A NATIONAL CONSERVATION STRATEGY FOR ZAMBIA

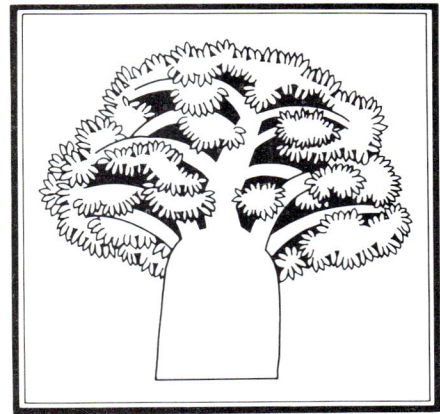

Conservation for Development

Conservation is by no means an idea which is entirely new to Zambia. But, until recently, there has been no all-encompassing management plan to make the most of its unusually abundant natural resources. While some resources are being destroyed or impaired, the utilization of others has long been far below potential.

Development must be balanced. It is wrong to obtain cheap electricity at the price of dried-up waterfalls, or short-term financial gain from uncontrolled tourism at the cost of the very sites the tourists have come to see.

It is in this context that the National Conservation Strategy for Zambia (NCSZ) has been formulated. The result of an intensive programme of consultations, it was prepared by the Ministry of Lands and Natural Resources, with technical assistance from IUCN and funding from the Swedish and Dutch Governments and WWF. Many other Government ministries, and organisations such as the Wildlife Conservation Society of Zambia, were also closely involved.

The NCSZ stems from a document known as the World Conservation Strategy (WCS). Prepared by IUCN, the United Nations Environment Programme and WWF in 1980, the WCS is based on the views of more than 700 scientists and 450 government agencies from over 100 countries. It shows the importance of sustainable development through rational utilization of living resources, identifies the priorities of conservation and presents an agreed approach and broad plan for achieving them.

To ensure that these objectives of conservation are achieved, the WCS recommends that every country reviews the extent to which it is accomplishing conservation within its own borders. The review is expected to form the basis of a national strategy which focuses attention on priority conservation requirements, stimulates appropriate action, raises public consciousness, and overcomes any apathy or resistance to taking the action needed.

When it was officially adopted by the Party and its Government in July 1985, the NCS for Zambia became the first such strategy to be completed in Africa.

Its long-term goal is to satisfy the basic needs of all the people of Zambia – both present and future generations – through the conservation and wise management of natural resources. Conservation is a widely-known practice in the wildlife field – and in many traditional agricultural practices – but it can be applied with equal justification and success to forests, water, industry, fisheries, pastures, energy and many other areas.

The Strategy works on the principle that conservation and development are two sides of the same coin. *Conservation can aid development* by managing the natural resources on which it depends. *Development can aid conservation* by ensuring that people are not obliged to over-exploit natural resources in order to survive.

But it is patently clear that no one discipline could have a monopoly of understanding about the environment and the way in which natural

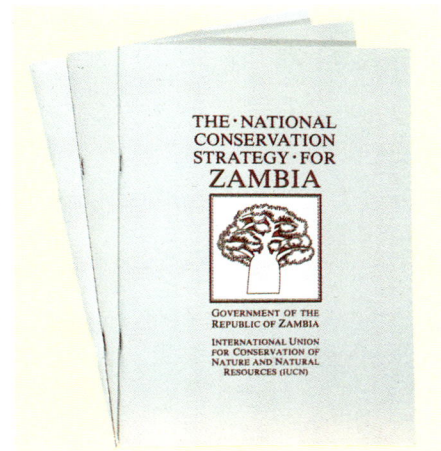

The National Conservation Strategy for Zambia became the first such strategy to be compiled in Africa.

Right: Strategy poster.

Below:
The theme of Zambia's National Conservation Strategy is 'making the most of what we have'. Its aim is to carefully manage the soil, water, forests, wild animal populations and other natural resources to ensure that they provide society's needs indefinitely.

ZAMBIA'S WATER RESOURCES ARE OUR FUTURE

resources can contribute to sustainable development. For the Strategy to work it requires the cooperation of politicians, economists, industrialists, engineers, farmers, biologists, teachers and many others. *All* Zambians use natural resources every day; therefore everyone has a responsibility for conserving them.

It should be a priority of every government to tackle environmental problems. But it is very easy to regard the long-term – and sometimes imperceptible – trends, characteristic of many environmental problems, as too remote to demand immediate attention.

Having recognised these problems and acknowledged that appropriate action needs to be taken, the task ahead now – to achieve balanced conservation and development in Zambia – is to implement the National Conservation Strategy. This work has already begun.

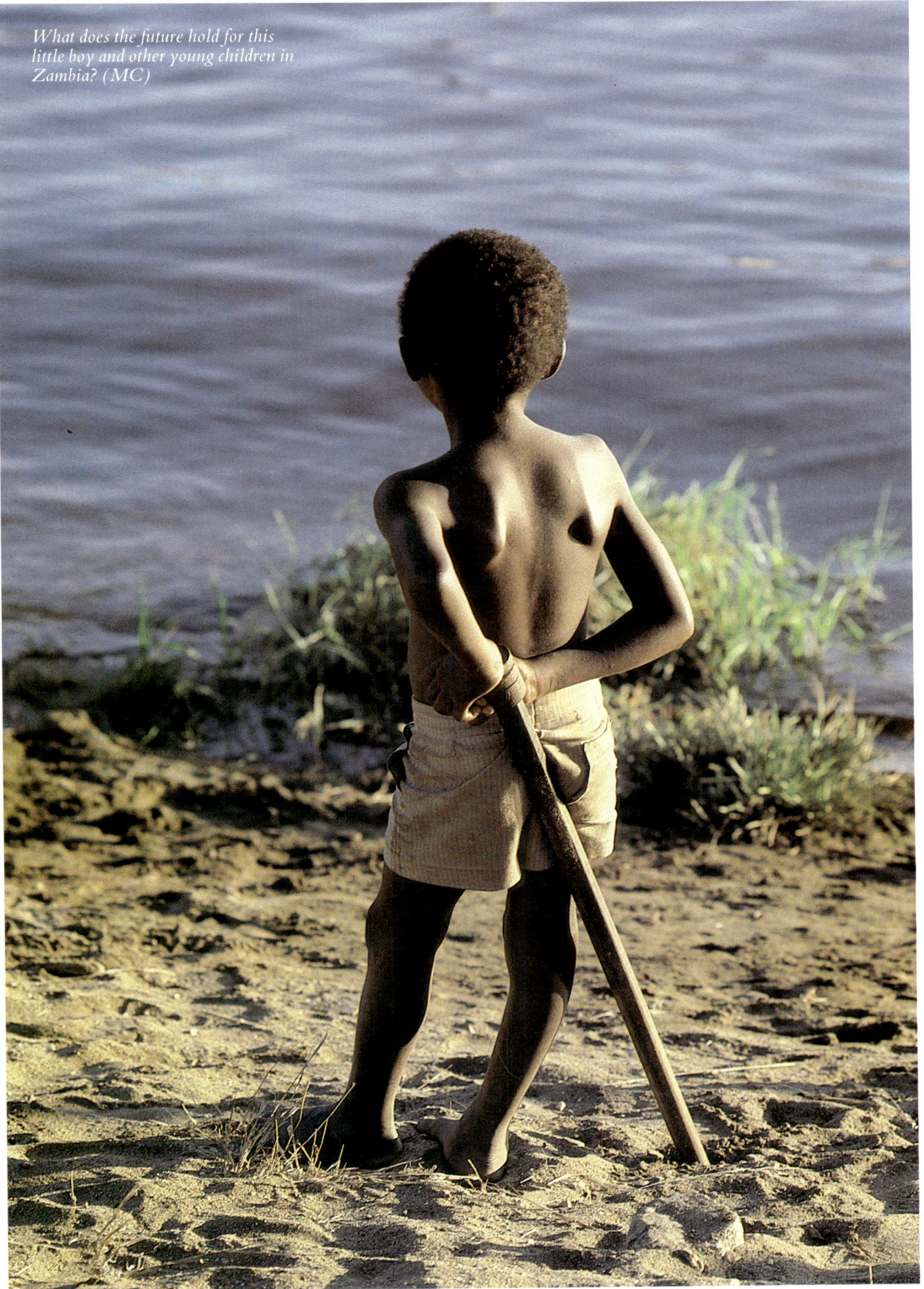

What does the future hold for this little boy and other young children in Zambia? (MC)

Acknowledgements

The enthusiastic response to all requests for assistance with the production of *The Nature of Zambia* was typical of the sympathy and motivation behind conservation efforts in Zambia. This particular project would have been impossible without the help, advice and support of so many people.

Financial assistance and services in kind came from a number of different sources. In particular, I would like to thank John A. Nkonde (Director of Marketing) and Mr. Muwowo (General Sales Manager) at Zambia Airways; Keith Green, Manager of the Pamodzi Hotel in Lusaka; and Jim Penman, Controller of Operations at the National Hotels Development Corporation.

The Honourable R.C. Kamanga, Member of the Central Committee and Chairman of the Rural Development Committee, and A.M. Sikota, Provincial Political Secretary (RDC), both of the United National Independence Party, were enormously helpful and very supportive.

A number of staff in the Ministry of Lands and Natural Resources helped from the beginning. It was, of course, largely their efforts which led to the production of Zambia's National Conservation Strategy, on which this publication is based. In particular, I would like to thank P. Mdala, Assistant Secretary, and Stanley Kashweka of the Forestry Department.

Among the many others who helped in different ways, particular thanks are due to Peter Juerges, Managing Director of Fleetfoot Advertising Ltd; E.M. Chidumayo, former Conservator of Natural Resources; Professor Geoff Howard, Chairman of the Kafue Basin Research Committee; Richard Lumbe and Mary Krag-Olsen at the Wildlife Conservation Society of Zambia; Norman Carr, world authority on Zambia's wildlife; Isaac Zulu, Iain Macdonald and Simon Bicknell of Chibembe Camp, South Luangwa; Richard Jeffery and Phil Berry of the SRT Chinzombo Safari Lodge, South Luangwa; Aldert van der Vinne and Tim Jones of Kashima and York Farms, respectively; Drs Kondanani Zulu and P.J. Mwanza, traditional healers; M.N. Katanekwa, Director, National Monuments Commission; H.M. Nyambe, Chief Postal Manager, Lusaka; Mike Faddy, Caleb Nkonga and the dedicated scouts of the Save the Rhino Trust; Gwenda Chongwe and Diana Fynn of Zintu Handicrafts; Alick Chanda, Director, Munda Wanga Gardens; Egbert Matibini, Production and Marketing Manager, Zambia National Tourist Board; Mr. Mbewe, Curator, Copperbelt Museum; the staff of the Livingstone Museum; and Fidelis Lungu, Co-Director, Luangwa Integrated Resource Development Project.

I would particularly like to thank Steve Bass, the Conservation for Development Centre's consultant in Zambia 1984-86, and his wife Christine, for their tremendous hospitality, advice and encouragement since the idea for *The Nature of Zambia* first began to take shape.

Photographer David Reed was a marvellous companion during much of the research. Most of the photography in this book is thanks to his outstanding skill and perseverance.

Mark Carwardine

*I*nterviewees

The Honourable R.C. Kamanga, Member of the Central Committee and Chairman of the Rural Development Committee

Alick Chanda, Curator of Munda Wanga Zoological and Botanical Gardens

M.N. Katanekwa, Director of the National Monuments Commission

Caleb Nkonga, Wildlife Warden, Save the Rhino Trust

Dr P.J. Mwanza, traditional healer and Provincial Secretary of the Traditional Health Practitioner's Association

Iain Macdonald, Manager of Chibembe Camp, South Luangwa

"KEEP ZAMBIA BEAUTIFUL"
SAMALANI ZAMBIA MWAUDONGO